Communications in Computer and Information Science 445

More information about this series at http://www.springer.com/series/7899

Clara Pizzuti · Giandomenico Spezzano (Eds.)

Advances in Artificial Life and Evolutionary Computation

9th Italian Workshop, WIVACE 2014
Vietri sul Mare, Italy, May 14–15
Revised Selected Papers

 Springer

Editors
Clara Pizzuti
Giandomenico Spezzano
CNR-ICAR
Rende
Italy

ISSN 1865-0929   ISSN 1865-0937 (electronic)
ISBN 978-3-319-12744-6   ISBN 978-3-319-12745-3 (eBook)
DOI 10.1007/978-3-319-12745-3

Library of Congress Control Number: 2014954577

Springer Cham Heidelberg New York Dordrecht London

Printed on acid-free paper

Springer is part of Springer Science+Business Media (www.springer.com)

Preface

This volume of the Springer book series Communications in Computer and Information Science contains the proceedings of WIVACE 2014, the Ninth Italian Workshop on Artificial Life and Evolutionary Computation, held from May 14 to 15, 2014, in Vietri Sul Mare, Italy.

The goal of WIVACE 2014 was to provide researchers in evolutionary computation, complex systems, and artificial life with an opportunity for the presentation of relevant novel researches in a strongly multidisciplinary context. Artificial Life and Evolutionary Computation (ALEC) are core research areas of what has become an exciting cross-fertilization between evolutionary biology, computer science, and engineering. Computer scientists and engineers, inspired by evolution in nature, realized that they could apply the same powerful Darwinian mechanism in computers for practical purposes, such as complex industrial design.

In recent years, there has been a market trend in the ALEC community toward real-world applications. Techniques, inspired by ALECs wide ambition to produce more intelligent systems, are not only gaining acceptance in other fields of scientific research, but also in areas such as business, commerce, and industry. The integration of these disciplines by different learning and adaptation techniques has in recent years contributed to the emergence of large numbers of new superior class of intelligence known as Hybrid Intelligence. Hybridization of different intelligent systems is an innovative approach to construct computationally intelligent systems consisting of artificial neural network, fuzzy inference systems, rough set, approximate reasoning, and optimization methods such as evolutionary computation, swarm intelligence, and particle swarm optimization.

To broaden its range, this year WIVACE 2014 was held in conjunction with the 24th Italian Workshop on Neural Networks WIRN 2014 (May 15–16). Participation in both workshops has been strongly encouraged in order to favor the interdisciplinary viewpoint of both communities.

The editors wish to express their sincere gratitude to all persons who supported this venture and made it feasible. In particular, we wish to thank all the authors who spent time and effort to contribute to this volume. We would also thank reviewers who, as members of the Program Committee, not only assessed papers, but also acted as session chairmen during the workshop. Special thanks, finally, to the invited speakers who, during the workshop gave three very interesting and inspiring talks: Enrique Alba, professor at University of Malága, Spain, Yaroslav D. Sergeyev, professor at University of Calabria, Italy, and Roberto Serra, professor at University of Modena and Reggio Emilia, Italy.

May 2014

Clara Pizzuti
Giandomenico Spezzano

Organization

WIVACE 2014 is organized by the Institute of High Performance Computing and Networking (ICAR) of National Research Council of Italy (CNR).

Program Chairs

Clara Pizzuti
    Institute of High Performance Computing and Networking (ICAR) of National Research Council of Italy (CNR), Italy

Giandomenico Spezzano
    Institute of High Performance Computing and Networking (ICAR) of National Research Council of Italy (CNR), Italy

Local Chairs

Alessia Amelio
    Institute of High Performance Computing and Networking (ICAR) of National Research Council of Italy (CNR), Italy

Andrea Giordano
    Institute of High Performance Computing and Networking (ICAR) of National Research Council of Italy (CNR), Italy

Andrea Vinci
    Institute of High Performance Computing and Networking (ICAR) of National Research Council of Italy (CNR), Italy

Program Committee

Alberto Acerbi	University of Bristol, UK
Marco Antoniotti	Università degli Studi di Milano-Bicocca, Italy
Michele Amoretti	Università degli Studi di Parma, Italy
Luca Ascari	Henesis Srl, Italy
Antonia Azzini	Università degli Studi di Milano, Italy
Lucia Ballerini	University of Edinburgh, UK
Armando Bazzani	Università di Bologna, Italy
Vitoantonio Bevilacqua	Politecnico di Bari, Italy
Leonardo Bocchi	Università degli Studi di Firenze, Italy
Andrea Bracciali	University of Stirling, UK
Ernesto Burattini	Università degli Studi di Napoli Federico II, Italy
Stefano Cagnoni	Università degli Studi di Parma, Italy
Raffaele Calabretta	ISTC-CNR, Italy

Angelo Cangelosi	Plymouth University, UK
Giulio Caravagna	Università degli Studi di Milano-Bicocca, Italy
Timoteo Carletti	University of Namur, Belgium
Antonio Chella	Università degli Studi di Palermo, Italy
Roberto Cordeschi	Università degli Studi di Roma "La Sapienza", Italy
Chiara Damiani	Università degli Studi di Milano-Bicocca, Italy
Giovanni De Matteis	Università degli Studi di Milano, Italy
Ivanoe De Falco	ICAR-CNR, Italy
Antonio Della Cioppa	Università degli Studi di Salerno, Italy
Giovanni De Matteis	Università degli Studi di Milano, Italy
Cecilia Di Chio	University of Southampton, UK
Marco Dorigo	IRIDIA, Université Libre de Bruxelles, Belgium
Alessandro Filisetti	Università di Bologna, Italy
Francesco Fontanella	Università degli Studi di Cassino e del Lazio Meridionale, Italy
Luigi Fortuna	University of Catania, Italy
Mario Giacobini	University of Turin, Italy
Alex Graudenzi	Università degli Studi di Milano-Bicocca, Italy
Marco Locatelli	Università degli Studi di Parma, Italy
Giancarlo Mauri	Università degli Studi di Milano-Bicocca, Italy
Elena Marchiori	Radboud University, The Netherlands
Orazio Miglino	Università degli Studi di Napoli Federico II, Italy
Marco Mirolli	ISTC-CNR, Italy
Alberto Moraglio	University of Birmingham, UK
Monica Mordonini	Università degli Studi di Parma, Italy
Luca Mussi	Università degli Studi di Parma, Italy
Giuseppe Nicosia	University of Catania, Italy
Stefano Nolfi	ICST-CNR, Italy
Pietro Pantano	University of Calabria, Italy
Mario Pavone	University of Catania, Italy
Stefano Pizzuti	ENEA, Italy
Riccardo Poli	University of Essex, UK
Simona E. Rombo	Università degli Studi di Palermo, Italy
Andrea Roli	Università di Bologna, Italy
Giuseppe Scollo	University of Catania, Italy
Roberto Serra	Università degli Studi di Modena e Reggio Emilia, Italy
Giovanni Squillero	Politecnico di Torino, Italy
Pietro Terna	University of Turin, Italy
Andrea Tettamanzi	University of Nice Sophia Antipolis, France
Vito Trianni	ISTC-CNR, Italy
Elio Tuci	Aberystwyth University, UK
Leonardo Vanneschi	University of Lisbon, Portugal
Marco Villani	Università degli Studi di Modena e Reggio Emilia, Italy

Sponsoring Institutions

Institute of High Performance Computing and Networking (ICAR)
National Research Council of Italy (CNR)
RES NOVAE Project - Reti Edifici Strade Nuovi Obiettivi Virtuosi per l'Ambiente e l'Energia

Contents

Building Energy Management Through Fault Detection Analysis Using Pattern Recognition Techniques Applied on Residual Neural Networks

Imran Khan[1], Alfonso Capozzoli[1], Fiorella Lauro[1,3(✉)],
Stefano Paolo Corgnati[1], and Stefano Pizzuti[2,3]

[1] Department of Energy (DENERG), Politecnico di Torino, Torino, Italy
[2] Italian National Agency for New Technologies,
Energy and Sustainable Economic Development (ENEA), Rome, Italy
[3] Computer Science and Automation Department, University Roma Tre, Roma, Italy
fiorella.lauro@polito.it

Abstract. In this paper a fault detection analysis through a neural networks ensembling approach and statistical pattern recognition techniques is presented. Abnormal consumption or faults are detected by analyzing the residual values, which are the difference between the expected and the real operating data. The residuals are more sensitive to faults and insensitive to noise. In this study, first, the experimentation is carried out over two months monitoring data set for the lighting energy consumption of an actual office building. Using a fault free data set for the training, an artificial neural networks ensemble (ANNE) is used for the estimation of hourly lighting energy consumption in normal operational conditions. The fault detection is performed through the analysis of the magnitude of residuals using peak outliers detection method. Second, the fault detection analysis is also carried out through statistical pattern recognition techniques on structured residuals of lighting power consumption considering different influencing attributes i.e. number of people, global solar radiation etc. Moreover the results obtained from these methods are compared to minimize the false anomalies and to improve the FDD process. Experimental results show the effectiveness of the ensembling approach in automatic detection of abnormal building lighting energy consumption. The results also indicate that statistical pattern recognition techniques applied to residuals are useful for detecting and isolating the faults as well as noise.

Keywords: Building · Energy · ANNs residuals · Pattern recognition · Fault detection

1 Introduction

Measuring and collecting large amount of data relevant to building energy consumption and overall performance is becoming increasingly available through

© Springer International Publishing Switzerland 2014
C. Pizzuti and G. Spezzano (Eds.): WIVACE 2014, CCIS 445, pp. 1–12, 2014.
DOI: 10.1007/978-3-319-12745-3_1

building energy management systems (BEMS). To make the best use of this big data, it is necessary to extract relevant information useful for energy optimization. In this paper different pattern recognition techniques i.e. classification and clustering are used for fault detection.

First, the fault detection is performed by evaluating the magnitude of the residuals generated by an artificial neural network ensemble (ANNE) using an outliers detection method (peak detection). The hourly energy consumption and maximum power for artificial lighting are monitored and used as targets (outputs) of the analyzed models. Second, the fault detection analysis is also performed through statistical pattern recognition techniques (CART, KMeans and DBSCAN) on structured residuals of peak power consumption for lighting considering different influencing attributes i.e. number of people, global solar radiation etc. In this research the capability of ANNE approach and effectiveness of statistical pattern recognition techniques using peak energy consumption residuals for artificial lighting fault detection of a real office building are investigated.

It should be noted that the fault detection of building lighting consumption is not a critical issue that requires a strictly real time execution. Thus, once the ANN models are trained and the ensemble and pattern recognition models are defined, all the simulations can be performed in minute order time. The proposed methodology allow to perform a fault detection analysis in "near" real time, i.e. with a shift of one hour, since an hourly timestamp was considered.

2   Motivation and Related Work

Buildings are one of the prime targets to reduce energy consumption around the world. Almost 32 % of the total energy consumption in industrialized countries is used for electricity, heating, ventilation, and air conditioning (HVAC) in buildings [8]. Furthermore, building industry is not only energy-intensive, but also knowledge-intensive. The real data of a building contains the actual information of building operation; and thus can reflect the building performance accurately [22]. For energy optimization, the evaluation of real time building energy consumption data is a demandable and emerging area of building energy analysis. Several studies have been published on methods for automatically detecting abnormal energy consumption data in buildings. Seem [19] presented pattern recognition with robust statistical outliers' detection method to investigate abnormal energy consumption. Liu et al. [12] used classification and regression tree method for whole building energy abnormal behavior. Some research works, [11,18,21], provided classification methods including the box plots approach, association rule mining and pattern recognition algorithm to detect anomalous energy consumption in buildings.

Yu et al. [22] used fuzzy neural networks model for fault detection and diagnosis on the energy consumption of the whole building. Fault-free measured data is used to build up the model and another measured data with a fault is used to validate the model and test the performance of fault detection. Model is applied on the measured data with fault of an open window in the room, and threshold for the fault detection is derived from the moving mean value and variance in a certain period (24 h). Dodier and Kreider [3] proposed neural network algorithm to evaluate whether the energy consumption data is normal. The energy consumption is predicted by collection of previous data by neural network. The ratio of actual energy consumption to expected energy consumption is calculated. The data is considered abnormal when the ratio is lower or higher than thresholds. Also, Holcomb et al. [6] proposed algorithmic techniques based on machine learning to address the prediction of building energy consumption from that of similar buildings in its geographical neighborhood and to localize the faults in building sub-systems.

3 Case Study and Data Introduction

The case study selected for the fault detection analysis is an office building located in Rome, Italy. The building is composed of three floors and is equipped with a monitoring system aimed at collecting energy consumption (electrical and thermal) and the environmental conditions. Moreover each room/office in the building is equipped with a presence sensor. In the paper, experiments are performed on a data set referred to energy consumption for artificial lighting only for the first floor. In this floor there are 13 offices and two CED rooms. Different number of fluorescent lamps (each 55 W) ranging from 4 to 8 are installed in each office/room. In the two CED rooms 12 lamps, each 55 W, are installed. In order to identify abnormal lighting energy consumption, the features considered as dependent variables for the models are the average hourly energy consumption and peak demand (maximum power). Both lighting energy and power consumption of buildings first floor are analyzed for the months of December and January. Furthermore, the independent variables that are recorded with an hourly timestamp are: people presence, number of active rooms (a room is considered active if at least one person is present), global solar radiation, time, date and day of the week. In order to verify the reliability and the effectiveness of the proposed methods two artificial faults have been created on 24th and 25th of January. In these days at the end of the working time with fewer people presence between 17:30 and 18:00 all artificial lights of the offices on the first floor have been switched on creating a peak of energy demand.

4 Brief Description of Proposed Methods

In this section a brief theoretical description of proposed methods is presented.

4.1 Consumption Modelling by Artificial Neural Network Basic Ensembling Method

Artificial neural networks (ANNs) [1,5] are black-box (or data-driven) models mainly used when analytical or transparent models cannot be applied to model complex relationships between inputs and outputs. The basic processing units of ANNs are neurons: the connections between neurons define the network topology or architecture. Among all the different types of interconnecting structures, the feedforward one is widely used: the data processing can extend over multiple (layers of) units, but no feedback connections are present, i.e., connections extending from outputs of units to inputs of units in the same layer or previous layers. These models are also known as multi-layer perceptrons (MLP) [17], since the basic structure is the perceptron [16].

An "ensemble" is a group of learning models working together on the same task to improve the performances of the constituent models. In the last years, several ensembling methods have been carried out [10,13]. The non-generative ensembling method seeks to combine the outputs of the models in the best way. In the case of ANNs, they are trained on the same data, they run together and their outputs are combined in a single one. In particular, basic ensemble method (BEM) [2,15], is the simplest non-generative ensembling method: it combines the outputs of M neural networks as their arithmetic mean.

4.2   Peak Detection Method and Mzscore

In many applications, such as building energy consumption analysis and savings, defining "peaks" in an objective way is very important for an easier identification in a given time-series. Thus, a peak can be defined as an observation that is inconsistent with the majority of observations of a data set.

The method considered in this work, peak detection method, calculates the value (score) of a peak function S for every element of the given time-series [14]. A given point is a peak if its score is positive and it is greater than or equal to a particular threshold value. Particularly, a peak function S computes the average of the maximum among the signed distances of a given point x_i in a time-series T from its k left neighbours and the maximum among the signed distances from its k right neighbours. The function S is an index that allows quantifying the severity of outliers and then provides information about the priorities for actions to be associated with each outlier. In addition to the function S, another synthetic index modified zscore ($Mzscore$) is used to determine the amount of variation from normal observations. This index is based on the distance and direction of each outlier compared to the average value of normal observations (observations that do not contain outliers).

4.3   Classification and Regression Tree (CART)

The CART algorithm is based on classification and regression trees. A CART is a binary decision tree that is constructed by splitting a parent node into two child nodes repeatedly, beginning with the root node that contains the whole learning sample. CART can easily handle both numerical and categorical variables and useful in robust detection of outliers. A decision tree is constructed from the recorded data which can easily be converted to classification rules for effective identification of anomalies. Therefore it is particularly suitable for conducting analysis of fault detection in real time. CART methodology generally consists of three parts: construction of maximum tree, choice of the right size tree and classification of new data [20].

4.4   Clustering

The selected algorithms can be classified into two categories: (i) partitioning methods and (ii) density-based methods. These methods require the definition of a metric to compute distances between objects in the dataset. In the case study analyzed, distances between objects are measured by means of the Euclidean distance computed on normalized data.

KMeans. It belongs to partitioning category [7], is able to find spherical-shaped clusters and is sensitive to the presence of outliers. It requires as input parameter k, the number of partitions in which the dataset should be divided. It represents each cluster with the mean value of the objects it aggregates, called centroid. The algorithm is based on an iterative procedure, preceded by a set-up phase, where k objects of the dataset are randomly chosen as the initial centroids. Each iteration performs two steps; in the first step, each object is assigned to the cluster whose centroid is the nearest to that object. In the second step centroids are relocated, by computing the mean of the objects within each cluster. Iterations continue until the k centroids do not change.

DBSCAN. It is a density-based method designed to deal with non-spherical shaped clusters and is less sensitive to the presence of outliers. DBSCAN [4] requires two input parameters, a real number r and an integer number $minPts$, used to define a density threshold in the data space. A high density area in the data space is an n-dimensional sphere with radius r which contains at least $minPts$ objects. DBSCAN is an iterative algorithm which iterates over the objects in the dataset, analyzing their neighborhood. The effectiveness of the algorithm is strongly affected by the setting of parameters r and $minPts$.

5   Results and Analysis

The ANN ensemble is built according to BEM, considering 10 feed-forward MLP ANNs, with 1 hidden layer consisting of 15 neurons, hyperbolic tangent as activation function for the hidden neurons, and linear for the output. The training period is approximately 4 weeks and testing period approximately 1 week. Simulations are performed with MATLAB R2012b through the Levenberg-Marquardt algorithm. The reported results (see Table 1) are averaged over the 10 different runs (standard deviation in brackets). Performance has been evaluated according to the mean absolute error (MAE) and the maximum absolute error (MAX):

$$MAE = \frac{1}{N} \sum_{i=1}^{N} |y_i - \hat{y}_i| \qquad (1)$$

Table 1. Experimental results (training and testing)

	TRAINING	ANN	BEM	TESTING	ANN	BEM
Active energy	MAE (kWh)	0.33(±0.01)	0.31	MAE (kWh)	0.66(±0.04)	0.63
	MAX (kWh)	1.41	1.06	MAX (kWh)	4.26	3.75
Maximum active power	MAE (kW)	0.35(±0.02)	0.33	MAE (kW)	0.81(±0.05)	0.78
	MAX (kW)	1.76	1.4	MAX (kW)	4.78	4.51

$$MAX = \max \left\{ |y_i - \hat{y}_i| \right\}_{i=1}^{N} \qquad (2)$$

where y_i is the real lighting consumption, $\hat{y}_i$ is the output of the model (estimated lighting consumption) and N is the size of the real data set.

As shown in Table 1, the results obtained with ANN BEM are slightly better than those obtained with constituent ANNs. In the following sections only the analysis performed on the maximum power for lighting is presented.

In order to estimate a normal pattern of the maximum electrical power for the artificial lighting, the training of the ANN BEM is performed considering a fault free data set, obtained through outlier detection. The lighting power demand is estimated really well through the ANN BEM in the training period. In the testing period the estimated power follow quite well the monitored power demand, with the exception of some evident abnormal values. The magnitude of the difference over the time between the actual and estimated power demand is analyzed for detecting anomalous situations. To this purpose the peak detection method has been applied to the residuals data set in the testing period. In Fig. 1 the trend of residuals over the time is shown and the abnormal detected power demand values are highlighted. The identified residual peaks include potential early morning faults, for which very high power demand is observed corresponding to the only cleaning staffs presence, and the two artificial faults. The results confirm that the analysis of residuals generated through the ANN BEM represents a useful and powerful technique for the peak building lighting fault detection.

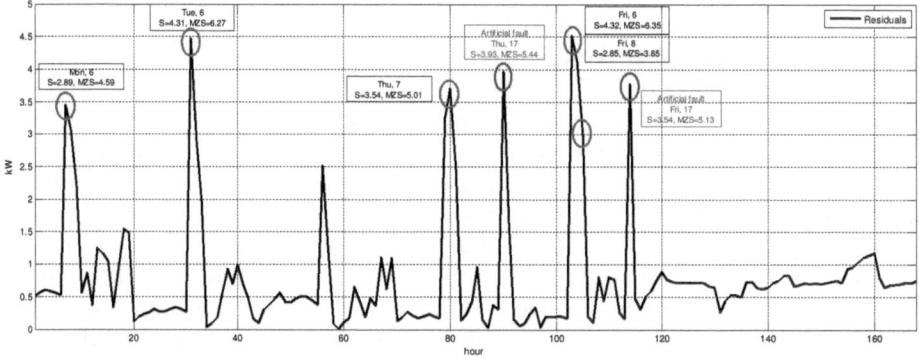

Fig. 1. Testing residuals (maximum active power) and detected peaks

Then, the peak detection method is applied to the maximum power consumption time-series. In Fig. 2 the outliers detected for testing period are shown with the relative values of Mzscore and S function indices. It can be observed that the method allows detecting the two artificial faults and some other real faults in early morning. In these situations the relative severity indices correctly assume higher value. However, the data show that power is related to other variables i.e. people, solar radiation, day and active rooms, so it can be inferred that the extreme values are not always definite faults. Therefore some false positives can be found when a univariate outlier detection method is applied without taking into account the effect of the independent variables on the consumptions.

In second part of this research, statistical pattern recognition techniques are applied on structured residuals of lighting power consumption. The major steps adopted for this analysis of fault detection are summarized as:

– Sensitivity analysis is carried out to identify the independent variable(s) of greater importance on the variation of the dependent variable (maximum power residuals), as shown in Fig. 3.
– CART is used for classification with one pruning method (number of cases in parent and child nodes). After multiple simulations and thorough analysis of the constructed classes, the number of cases set for parent and child nodes are 40 and 20 respectively. The independent variables considered for the classification are day, date, time, people presence, active rooms and solar radiations. The data are divided into 4 classes and each class has been analyzed separately. The classes are formed with time as most influential factor which is also evident from Fig. 3. In Fig. 4a and b scatter plots for class 1 are shown highlighting the two artificial faults. The class 1 contains the data values of early

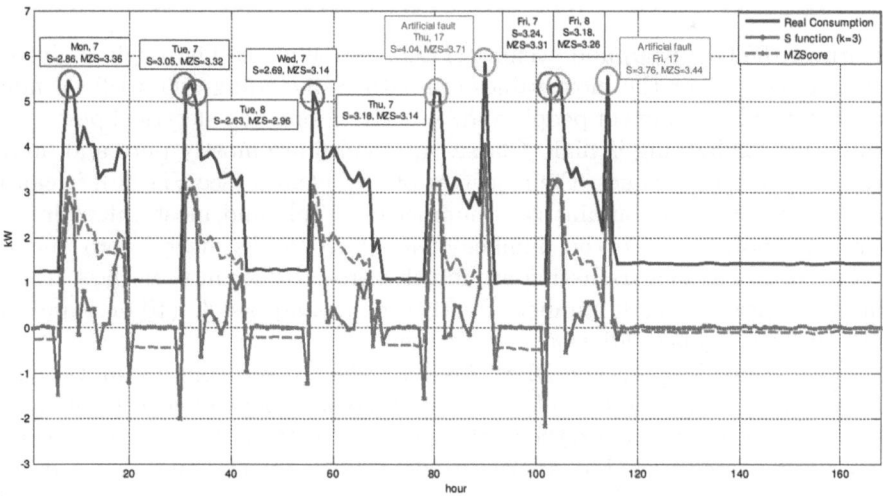

Fig. 2. Maximum active power (testing period), S function values, mzscore and detected peaks (common peaks are orange) (Color figure online)

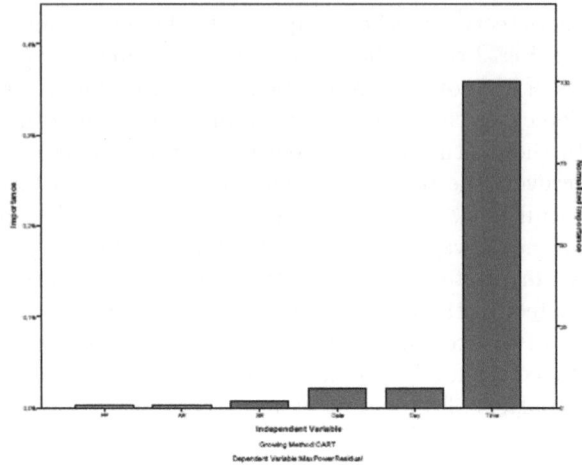

Fig. 3. Sensitivity analysis

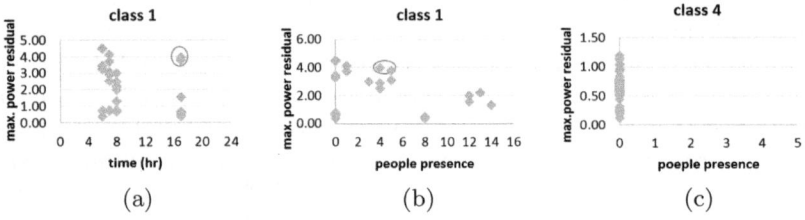

(a)   (b)   (c)

Fig. 4. Scatter plots for class 1 (CART) with artificial outliers encircled and for class 4

morning (06:00–08:00) and evening (17:00) for all week. Thorough analysis of class 1 shows that the most values are outliers including two artificial faults as with fewer numbers of people and/or active rooms the electrical power consumption for lighting is high. Classes 4, 5 and 6 are mostly pure and do not contain abnormal values. From scatter plot of class 4 (see Fig. 4c), it can be seen that the class contains zero number of people and most values in that class are normal. Also the number of active rooms is always zero and the values of global solar radiation are mostly zero or fewer in that class. Class 6 mostly consists of high number of active rooms and people (10 or more) and high values of solar radiation.

– For clustering (KMeans and DBSCAN), in order to overcome the limitations of the algorithms that do not allow time and day as independent variables, data sets have been divided into the working period (07:00–18:00), the non-working period and weekends. The approach adopted for the splitting of the data is the experience gained from our previous work [9] for which the division of the data set in the daytime, nighttime and weekend proved not to be particularly effective for the nature of the fault in the type of building under investigation.

It did not allow effectively the detection of outliers present in the early hours of the morning and at the end of working hours. Also before performing the clustering analysis, values of both dependent (max. power residuals) and independent variables (people presence, active rooms and solar radiations) have been normalized by means of standard score (z-score) method. For working, non-working and weekends the data is divided into 3, 2 and 2 clusters respectively using KMeans clustering. For DBSCAN clustering all three different split data sets are divided into two clusters each. To set the input parameters $(r, minPts)$ multiple tests are carried out for each data set by using different values for these parameters. With DBSCAN method, in all discovered clusters, the cluster label zero contains all points identified as outliers or noise.

Tables 2 and 3 show the cluster 1 (KMeans) and cluster zero (DBSCAN) clustering respectively performed on working hours data set. The results show that the splitting criteria used in this research has proved effective in overcoming the limitations of cluster algorithms and produced good results. Both the clusters are impure and contain artificial outliers with other positive outliers too. Clusters 2 and 3 (KMeans-working hours) are pure. In Fig. 5a and b scatter plots for cluster 2 (KMeans) are given. The cluster includes higher number of people and active rooms and energy consumption can be considered as normal.

For non-working hours, both cluster 2 (KMeans) and zero cluster (DBSCAN) have higher values of energy consumption corresponding to zero number of people presence in the early morning (06:00 am). For weekends, the clusters are formed

Table 2. KMeans (working hours, cluster 1), artificial faults are highlighted

Day	Date	Hour	Maximum power residual	Maximum power (kW)	PP	AR	SR	Clu-1
Mon	21/01/2013	7	3.07	5.45	5	5	0.83	1
Tue	22/01/2013	7	2.88	5.41	4	4	3.80	1
Wed	23/01/2013	7	2.51	5.2	4	4	7.00	1
Thu	24/01/2013	7	3.70	5.2	1	1	2.00	1
Thu	24/01/2013	8	2.47	5.19	4	3	27.00	1
Thu	**24/01/2013**	**17**	**3.96**	**5.86**	**4**	**3**	**2.67**	**1**
Thu	24/01/2013	18	1.64	3.08	3	2	0.00	1
Fri	25/01/2013	7	4.12	5.39	1	1	2.50	1
Fri	25/01/2013	8	3.00	5.33	3	3	19.62	1
Fri	**25/01/2013**	**17**	**3.77**	**5.55**	**5**	**4**	**1.75**	**1**

Table 3. DBSCAN (working hours, cluster 0), artificial faults are highlighted

Day	Date	Hour	Maximum power residual	Maximum power (kW)	PP	AR	SR	DBSCAN
Thu	24/01/2013	7	3.70	5.20	1	1	2.00	cluster_0
Thu	24/01/2013	12	0.96	3.27	6	5	476.45	cluster_0
Thu	**24/01/2013**	**17**	**3.96**	**5.86**	**4**	**3**	**2.67**	**cluster_0**
Fri	25/01/2013	7	4.12	5.39	1	1	2.50	cluster_0
Fri	25/01/2013	8	3.00	5.33	3	3	19.62	cluster_0
Fri	**25/01/2013**	**17**	**3.77**	**5.55**	**5**	**4**	**1.75**	**cluster_0**

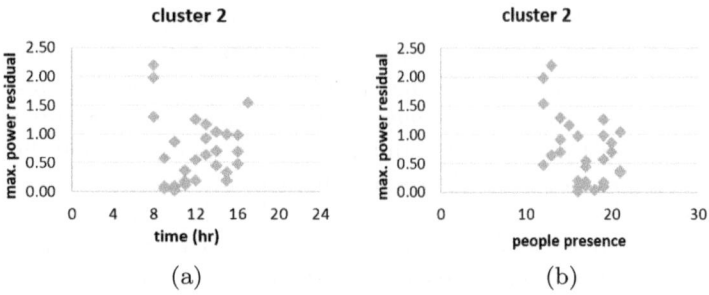

Fig. 5. Scatter plots for cluster 2 (KMeans-working hours)

Table 4. Some common outliers in all methods

Day	Date	Hour	Maximum power residual	Maximum power (kW)	PP	AR	SR
Mon	21/01/2013	6	3.45	4.17	0	0	0.00
Mon	21/01/2013	7	3.07	5.45	5	5	0.83
Mon	21/01/2013	8	2.20	5.23	13	9	19.75
Tue	22/01/2013	6	4.46	5.26	0	0	0.00
Tue	22/01/2013	7	2.88	5.41	4	4	3.80
Wed	23/01/2013	7	2.51	5.2	4	4	7.00
Thu	24/01/2013	6	3.26	4.2	0	0	0.00
Thu	24/01/2013	7	3.70	5.2	1	1	2.00
Thu	24/01/2013	8	2.47	5.19	4	3	27.00
Thu	**24/01/2013**	**17**	**3.96**	**5.86**	**4**	**3**	**2.67**
Fri	25/01/2013	6	4.51	5.34	0	0	0.00
Fri	25/01/2013	7	4.12	5.39	1	1	2.50
Fri	25/01/2013	8	3.00	5.33	3	3	19.62
Fri	**25/01/2013**	**17**	**3.77**	**5.55**	**5**	**4**	**1.75**

with solar radiation as the most influential factor. Also the outliers detected by each method are compared and some common outliers are presented in Table 4. By analyzing the results obtained from each method, it can be concluded that, in general, outliers are identified in two different periods of the day. The first period is early morning (06:00–08:00 am). In the early morning electrical power for lighting has the peaks at a very low presence of occupants. The second period is related to the end of working hours (17:00), where it is observed that a decrease in the number of occupants of the building does not correspond to a decrease in electrical power consumption for lighting.

6   Conclusions

To achieve objectives for energy efficiency, it is necessary to evaluate information contained in sensed building data. This paper presented a new approach by combining the ANNE with statistical pattern recognition techniques for evaluating real energy consumption data for whole building lighting energy fault detection. The research is aimed at investigating the potential of using ensembling techniques and, additionally, usefulness of statistical pattern recognition methods performed on structured residuals for fault detection.

The fault detection, performed through the analysis of the magnitude of residuals using a peak detection method, allowed to detect the two artificial faults and some other actual anomalous power values in the testing data set. Finally, the results obtained through all statistical pattern recognition techniques, performed on structured residuals, proved to be adequate and each method has been able to detect artificial faults and other positive outliers.

The application of this approach can improve fault detection process by reducing the number of false anomalies. The data set considered in this study is relatively small and only for artificial lighting of single building. In future, the work will consider new end-uses i.e. HVAC, plug load and exploit other data mining techniques for fault detection.

References

1. Arbib, M.A.: The Handbook of Brain Theory and Neural Networks. MIT press, Cambridge (2003)
2. Bishop, C.M.: Neural Networks for Pattern Recognition. Oxford University Press Inc, New York (1995)
3. Dodier, R.H., Kreider, J.F.: Symposium papers-ch-99-5-fault detection and diagnostics-learning from building operations-detecting whole building energy problems. ASHRAE Trans. Am. Soc. Heating Refrig. Airconditioning Engin **105**(1), 579–592 (1999)
4. Ester, M., Kriegel, H.P., Sander, J., Xu, X.: A density-based algorithm for discovering clusters in large spatial databases with noise. In: KDD, vol. 96, pp. 226–231 (1996)
5. Haykin, S.: Neural Networks: A Comprehensive Foundation, 2nd edn. Prentice Hall PTR, Upper Saddle River (1998)
6. Holcomb, D., Li, W., Seshia, S.A.: Algorithms for green buildings: learning-based techniques for energy prediction and fault diagnosis. Google Scholar, UCB/EECS-2009-138 (2009)
7. Juang, B.H., Rabiner, L.: The segmental k-means algorithm for estimating parameters of hidden markov models. IEEE Trans. Acoust. Speech Sig. Process. **38**(9), 1639–1641 (1990)
8. Katipamula, S., Brambley, M.R.: Review article: methods for fault detection, diagnostics, and prognostics for building systems a review, part i. HVAC&R Res. **11**(1), 3–25 (2005)
9. Khan, I., Capozzoli, A., Corgnati, S.P., Cerquitelli, T.: Fault detection analysis of building energy consumption using data mining techniques. Energy Procedia **42**, 557–566 (2013)
10. Krogh, A., Vedelsby, J., et al.: Neural network ensembles, cross validation, and active learning. In: Tesauro, G. (ed.) Advances in Neural Information Processing Systems, pp. 231–238. MIT Press, Cambridge (1995)
11. Li, X., Bowers, C.P., Schnier, T.: Classification of energy consumption in buildings with outlier detection. IEEE Trans. Ind. Electron. **57**(11), 3639–3644 (2010)

12. Liu, D., Chen, Q., Mori, K., Kida, Y.: A method for detecting abnormal electricity energy consumption in buildings. J. Comput. Inf. Syst. **6**(14), 4887–4895 (2010)
13. Liu, Y., Yao, X.: Ensemble learning via negative correlation. Neural Netw. **12**(10), 1399–1404 (1999)
14. Palshikar, G., et al.: Simple algorithms for peak detection in time-series. In: Proceedings of the 1st International Conference Advanced Data Analysis, Business Analytics and Intelligence (2009)
15. Perrone, M.P., Cooper, L.N.: When networks disagree: ensemble methods for hybrid neural networks. Technical report, DTIC Document (1992)
16. Rosenblatt, F.: The perceptron-a perceiving and recognizing automaton. Technical report 85–460-1, Cornell Aeronautical Laboratory, Buffalo, NY, 00344 (1957)
17. Rosenblatt, F.: Principles of neurodynamics. Perceptrons and the theory of brain mechanisms. Technical report, DTIC Document (1961)
18. Seem, J.E.: Pattern recognition algorithm for determining days of the week with similar energy consumption profiles. Energy Buildings **37**(2), 127–139 (2005). http://www.sciencedirect.com/science/article/pii/S0378778804001434
19. Seem, J.E.: Using intelligent data analysis to detect abnormal energy consumption in buildings. Energy Buildings **39**(1), 52–58 (2007). http://www.sciencedirect.com/science/article/pii/S0378778806001514HrB
20. Timofeev, R.: Classification and regression trees (cart) theory and applications (2004)
21. Yoshida, K., Inui, M., Yairi, T., Machida, K., Shioya, M., Masukawa, Y.: Identification of causal variables for building energy fault detection by semi-supervised lda and decision boundary analysis. In: 2008 IEEE International Conference on Data Mining Workshops, ICDMW'08, pp. 164–173. IEEE (2008)
22. Yu, B., van Paassen, D.H.: Fuzzy neural networks model for building energy diagnosis. In: 8th International IBPSA Conference, pp. 1459–1466 (2003)

Qualitative Particle Swarm Optimization (Q-PSO) for Energy-Efficient Building Designs

Debora Slanzi[1,2](✉), Matteo Borrotti[1,2], Davide De March[1,2],
Daniele Orlando[1], Silvio Giove[3], and Irene Poli[1,2]

[1] European Centre for Living Technology, S. Marco 2940, 30124 Venice, Italy
debora.slanzi@unive.it
[2] Department of Environmental Sciences, Informatics and Statistics,
Ca' Foscari University of Venice, Cannaregio 873, 30121 Venice, Italy
[3] Department of Economics, Ca' Foscari University of Venice,
Cannaregio 873, 30121 Venice, Italy

Abstract. Particle Swarm Optimization (PSO) is a stochastic optimization method, based on the social behavior of bird flocks. The method, known for its high performance in optimization, has been mainly developed for problems involving just quantitative variables. In this paper we propose a new approach called Qualitative Particle Swarm Optimization (Q-PSO) where the variables in the optimization can be both qualitative and quantitative and the updating rule is derived adopting probabilistic measures. We apply this procedure to a complex engineering optimization problem concerning building façade design. More specifically, we address the problem of deriving an energy-efficient building design, i.e. a design that minimizes the energy consumption (and the emission of carbon dioxide) for heating, cooling and lighting. We develop a simulation study to evaluate Q-PSO procedure and we derive comparisons with most conventional approaches. The study shows a very good performance of our approach in achieving the assigned target.

Keywords: Qualitative particle swarm optimization · Engineering optimization · Energy-efficient building design

1 Introduction

Particle Swarm Optimization (PSO) is an optimization method inspired by the social behavior of bird flocks or fish schools. Introduced by Kennedy and Eberhart [1] as a stochastic optimization algorithm, PSO is a population-based search procedure in which the population is conceived as a swarm composed of particles. In this approach each particle moves in the search space with an adaptable velocity, recording the best position it has ever visited in the search space, i.e. the position with the lowest objective function value when we deal with the minimization problems. Each particle has a neighborhood that consists of some pre-defined particles and the best position attained so far by each

C. Pizzuti and G. Spezzano (Eds.): WIVACE 2014, CCIS 445, pp. 13–25, 2014.
DOI: 10.1007/978-3-319-12745-3_2

member of the neighborhood is communicated to the particle itself and affects its movement. In this way a particle moves in the search space looking for the optimal values with both the information on its best position, called *cognitive component*, and the information on the best position reached by the neighborhood, called *social component*. PSO has been applied to solve a large number of optimization problems [2], mainly to search in continuous domains. There are few variants of the approach that operate in discrete spaces, such as the procedures developed for binary problems [3], or integer and combinatorial problems [4, 6].

In this paper we propose an innovative PSO approach able to deal with problems charesterized by both qualitative and quantitative variables. We develop this approach to address an optimization problem for the design of energy-efficient building envelopes. More specifically we consider the problem of minimizing the carbon dioxide emissions due to the heating, cooling and lighting energy consumption in a room regarded as a module of a building.

The paper is structured as follows. In Sect. 2 we describe PSO approach and we show the updating rules by which the particles change their position in the search space; in Sect. 3 we introduce the *Qualitative Particle Swarm Optimization* (Q-PSO) to address problems that involve qualitative variables. In Sect. 4 we perform a simulation study to test the approach and evaluate its performance in deriving optimal values for designing energy-efficient building façades. Finally in Sect. 5, we derive some remarks on the performance of the approach also in comparison with other procedures.

2   Particle Swarm Optimization (PSO) Approach

Addressing continuous optimization problems which involve the optimization of an objective function $h(\mathbf{x})$ of a variable vector $\mathbf{x}$ defined on a D-dimensional space, PSO algorithm adopts a population (swarm) of particles that adjust their position in time according to their own information and neighborhood particles information. In this procedure the swarm S is composed by P *particles*. At each iteration (step t of the algorithm), the i-th particle of the swarm is associated with the position in the continuos D-dimensional search space $\mathbf{x}_i(t) = [x_{i,1}(t), x_{i,2}(t), \ldots, x_{i,D}(t)]$, and with its velocity, describing the last particle position change, $\mathbf{v}_i(t) = [v_{i,1}(t), v_{i,2}(t), \ldots, v_{i,D}(t)]$. To the i-th particle it is then associated the objective function $h(\mathbf{x}_i)$, with $i \in \{1, 2, \ldots, P\}$.

At the first step of the procedure, both the position and velocity vectors are randomly initialized within a range of feasible values. Then, both of these parameters are iteratively updated until a stopping criterion is met. The update equation rules are:

$$v_{i,d}(t+1) = f\left(v_{i,d}(t), x_{i,d}(t), pbest_{i,d}(t), nbest_{i,d}(t)\right) \tag{2.1}$$

$$x_{i,d}(t+1) = g\left(x_{i,d}(t), v_{i,d}(t+1)\right) \tag{2.2}$$

where $f(\cdot)$ and $g(\cdot)$ are suitable functions, $pbest_{i,d}(t)$ is the d-th element of the vector $\boldsymbol{pbest}_i(t)$ representing the historical best position of the particle i, i.e.

the lowest objective function value (personal best), and $nbest_{i,d}(t)$ is the d-th element of vector $\boldsymbol{nbest}_i(t)$ related to the best position of the i-th particle's neighborhood (social best).

According to these rules, the i-th particle position at step $(t+1)$ is determined by the function $g(\cdot)$ of its previous position and its velocity at time t. Moreover the velocity of i-th particle at time $(t+1)$ is a function $f(\cdot)$ depending on the velocity and the position at time t, and the best positions achieved by the particle and its neighborhood. Based on the paper of Kennedy and Eberhart [1], the functions $f(\cdot)$ and $g(\cdot)$ are specified by the following equations:

$$
\begin{aligned}
v_{i,d}(t+1) =\ & v_{i,d}(t) + c_1\rho_1\left(pbest_{i,d}(t) - x_{i,d}(t)\right) + \\
& c_2\rho_2\left(nbest_{i,d}(t) - x_{i,d}(t)\right) \qquad\qquad (2.3)\\
x_{i,d}(t+1) =\ & x_{i,d}(t) + v_{i,d}(t+1) \qquad\qquad\qquad\qquad\ \ (2.4)
\end{aligned}
$$

where c_1 and c_2 are two positive constants representing the cognitive and social acceleration coefficients; $\rho_1, \rho_2 \sim \mathcal{U}(0,1)$ are two independent uniformly distributed random values in the range $[0,1]$ introduced to weight the velocity toward the particle personal best and the velocity toward the global best solution.

When the neighborhood of each particle is represented by the whole swarm, then $\boldsymbol{nbest}_i$ naturally becomes $\boldsymbol{gbest}$, that is the global best solution. This condition is known as the *gbest topology*. Under this condition all the particles are connected among them, achieving a full connected graph: the unique best particle position in the entire population affects all the other particles positions.

One of the most problematic characteristic of PSO is its propensity to converge, prematurely, on early best solutions. In the paper of Shi and Eberhart [7] a *inertia* weight, w, was introduced to control the velocity of the particles in order to overcome this limitation. In their successive paper [8], a linearly decreasing w is proposed changing its value according to the number of iterations. Further, in order to improve the performance of the method, Chatterrjee and Siarry [9] suggested an updating equation for the *inertia* weight w that introduces a non linear component in the velocity function. Rather then considering the *inertia* weight, Clerc and Kennedy [5] introduced the constriction coefficient, which alleviates the requirement to clamp the velocity.

Considering discrete optimization problems, several PSO algorithms have been proposed in the literature. Kennedy and Eberhart [3] introduced a procedure able to deal with sequences of bits rather than real numbers. In this approach each element of the velocity vector is bound in the interval $[0,1]$ through a sigmoid function, setting a probability threshold to flip the d-th element of the vector $\mathbf{x}_i(t)$ from 0 to 1 or vice versa. This procedure can also be adopted when the accessible values are more than two, simply converting decimal to binary numbers (i.e. gray code). Liao et al. [10] extended discrete PSO [3] to solve the flow-shop scheduling problem by redefining the particle and the velocity and incorporating a local search scheme to move a particle to the new sequence. Dealing with integer problems, the approach proposed by Laskari et al. [4] addresses the issue by simply rounding the updated value $x_{i,d}(t+1)$ in (2.4) to the nearest integer $x^*_{i,d}(t+1)$. For a review on PSO for discrete optimization problems see [11].

Addressing directly the problem of optimization involving both qualitative and quantitative variables, we propose a novel approach based on the concept of probabilistic attraction.

3   Qualitative Particle Swarm Optimization (Q-PSO)

Many real-world optimization problems require the involvement of qualitative variables that are variables whose values are represented by a finite set of labels. In this work, we consider nominal qualitative variables, i.e. variables described by non-ordered labels.

In order to derive an optimization procedure based on the fundamental principles of PSO and suitable to deal with qualitative variables, we propose to introduce in the algorithm an updating probabilistic rule. More specifically we propose to update the value of a qualitative variable using a probability distribution rather than a velocity parameter. For a particle i, we specify the position vector as follows:

$$\mathbf{x}_i = (\underbrace{x_{i,1}, x_{i,2}, \ldots, x_{i,Q}}_{\text{qualitative variables}}, \underbrace{x_{i,Q+1}, x_{i,Q+2}, \ldots, x_{i,Q+C}}_{\text{quantitative variables}}), \tag{3.1}$$

$$\mathbf{x}_i \in L_1 \times \ldots \times L_Q \times \mathbb{R}^C$$

where the first Q variables are qualitative while the remaining C variables are quantitative (continuous). We then denote with L_q the set of labels of the q-th qualitative variable, with $q \in \{1, 2, \ldots, Q\}$ and n_{L_q} the cardinality of the set L_q.

The position and the velocity of the quantitative variables can be updated using the rules introduced by canonical PSO algorithms. The parameters of the qualitative variables introduced in this algorithm require instead the specification of a new set of equations.

To build a PSO algorithm that allows the consideration of qualitative variables, we introduce the concept of *probabilistic attraction*. The probabilistic attraction consists in sampling from the L_q possible labels of each qualitative variable with a probability distribution. At generation $t + 1$, this distribution depends on the contribution of each label in determining the global and local best positions at generation t of the algorithm. In this way we sample more frequently those labels that contribute most to determine optimal values of the objective function, and less frequently the others. With the probabilistic attraction procedure we also assign a non-zero probability to choose other labels of the variable, avoiding in this way to get stuck in some local optimum.

Considering the i-th particle, with $i \in \{1, 2, \ldots, P\}$, and the q-th qualitative variable, with $q \in \{1, 2, \ldots, Q\}$, at generation $t + 1$ we define the probabilistic attraction procedure as follows.

Let π be the equal sampling probability of each label of the q-th variable (equiprobability condition), $_i\pi_q^{pbest}(t)$ the probability to choose the individual best position of particle i for the q-th variable at generation t, $\mathbf{X}(t)$ the position matrix of all the particles in the swarm at time t, $\mathbf{X}(t) = [\mathbf{x}_1(t), \mathbf{x}_2(t), \ldots, \mathbf{x}_P(t)]^T$, and $\mathbf{\Pi}(t)$ the matrix of the individual best positions of each particle in the swarm at generation t, $\mathbf{\Pi}(t) = [\boldsymbol{pbest}_1(t), \boldsymbol{pbest}_2(t), \ldots, \boldsymbol{pbest}_P(t)]^T$.

Then the sampling probabilities of each label in L_q, based on the results obtained at the current generation t, is determined by:

1. the identification of the individual best label $_iL_q^{pbest}(t)$ of the particle i at the current generation t, with respect to the objective function to be optimized;
2. the identification of the best global label $L_q^{gbest}(t)$ of the whole swarm at the current generation t, in optimizing the objective function;
3. the assessment of the probability $_i\pi_q^{pbest}(t+1)$ to choose the individual best label $_iL_q^{pbest}(t)$ at the next generation $t+1$

$$_i\pi_q^{pbest}(t+1) = \Pr\{x_{i,q}(t+1) = {}_iL_q^{pbest}(t)\}$$
$$= f\left(\pi, {}_i\pi_q^{pbest}(t), \mathbf{X}(t), \mathbf{\Pi}(t)\right), \quad (3.2)$$

where $f(\cdot)$ is a suitable function;
4. the computation of the probability $\pi_q^{gbest}(t+1)$ to choose the global best label $L_q^{gbest}(t)$ at the next generation $t+1$

$$\pi_q^{gbest}(t+1) = \Pr\{x_{i,q}(t+1) = L_q^{gbest}(t)\}$$
$$= g\left(\pi, \pi_q^{gbest}(t), \mathbf{X}(t), \mathbf{\Pi}(t)\right), \quad (3.3)$$

where $g(\cdot)$ is a suitable function and $_i\pi_q^{pbest}(t)$ in 3.2 is substituted by $\pi_q^{gbest}(t)$.
5. the computation of the probabilities of the labels that are not the individual or global best as follows:

$$_i\pi_q^l(t+1) = h\left(\pi_q^{gbest}(t+1), {}_i\pi_q^{pbest_i}(t+1), n_{L_q}\right),$$
$$\forall l \in L_q, \ l \notin \{L_q^{gbest}, {}_iL_q^{pbest}\}. \quad (3.4)$$

where $h(\cdot)$ is a suitable function.

In this paper the functions in (3.2), (3.3) and (3.4) assume fixed values according to the case study that we will address. In particular we assume that $h(\cdot)$ depends on π_q^{gbest} and $_i\pi_q^{pbest}$, as described in the following equation:

$$_i\pi_q^l(t+1) = \frac{_i\pi_q^{res}(t+1)}{(n_{L_q}-2)}, \quad \forall l \in L_q, \ l \notin \{L_q^{gbest}, {}_iL_q^{pbest}\}, \quad (3.5)$$

where

$$_i\pi_q^{res}(t+1) = 1 - \left[\pi_q^{gbest}(t+1) + {}_i\pi_q^{pbest}(t+1)\right]. \quad (3.6)$$

For the k-th particle, which defines the best global label of the whole swarm, we have $L_q^{gbest} = {}_kL_q^{pbest} = L_q^{best}$. In this case we define its distribution probability as a monotonically increasing function $\zeta(\cdot)$ as follows:

$$\Pr\{x_{k,q}(t+1) = L_q^{best}(t)\} = {}_k\pi_q^{best}(t+1) = \zeta\left(\pi_q^{gbest}, {}_k\pi_q^{pbest}\right). \quad (3.7)$$

This situation happens at least one time in each generation of the algorithm, because the global best position is selected among the individual best positions. In particular we assume that $_k\pi_q^{best}(t+1) = \left[\pi_q^{gbest}(t+1) + {}_k\pi_q^{pbest}(t+1)\right]$.

According to this assumption, the probability of each label different from L_q^{best} could be derived as in Eq. (3.5), and normalized to guarantee the constraint $\sum_{j=1}^{n_L} \pi_q^j(t+1) = 1$. Then the probability of sampling one of the remaining labels is calculated as follows:

$$_k^*\pi_q^l(t+1) = \frac{{}_k\pi_q^l(t+1)}{\sum_{j=1}^{n_L} {}_k\pi_q^j}, \quad l \in \{1, \ldots, n_L\}. \tag{3.8}$$

At time $t+1$, when all the probability values are determined, we update the q-th qualitative variable of $\mathbf{x}_i$ according to the following distribution function:

$$x_{i,q} = \begin{cases} L_q^{gbest}(t) & \pi_q^{gbest}(t+1) \\ {}_iL_q^{pbest}(t) & {}_i\pi_q^{pbest}(t+1) \\ {}_iL_q^l \neq \{L_q^{gbest}(t), {}_iL_q^{pbest}(t)\} & {}_i\pi_q^l(t+1) \quad {}_{l\notin\{pbest,gbest\}}, \end{cases} \tag{3.9}$$

where $\pi_q^{gbest}(t+1) > {}_i\pi_q^{pbest}(t+1) > {}_i\pi_q^l(t+1)$.

We denote this approach *Qualitative Particle Swarm Optimization (Q-PSO)*. It represents a generalization of the particle swarm optimization approach, since allows framing the problems considering both quantitative and qualitative variables. In this paper we will derive Q-PSO approach for the problem of designing building envelopes with the objective of reducing the consumption of energy. Some of the variables involved in this optimization problem are in fact qualitative, and a more general PSO approach may lead to a more accurate analysis.

4   Q-PSO for Energy-Efficient Building Envelope Designs

The reduction of pollutant emissions is nowadays a fundamental problem related to the climate change issue and environmental sustainability. Major causes of air pollution are the emission of carbon dioxide from energy building consumption, vehicular traffic, and industrial production. In this article, we focus on the objective to reduce building carbon dioxide emissions meanwhile maintaining living and comfort conditions inside buildings. The quantity of energy consumed for heating, cooling and lighting a building is a function of several variables (qualitative and quantitative) and parameters, including the climatic conditions of the construction site, the building orientation, the materials, the type of insulation and the geometry of the interior and exterior. In this field, several studies have addressed the problem by using different optimization procedures, which include the evolutionary optimization approaches [12–14]. In particular, Zemella et al. [13] proposed an Evolutionary Neural Network to optimize a building envelope design where some qualitative variables were involved.

In designing energy-efficient building façades several variables have to be considered and their single and interactive effects on energy consumption should

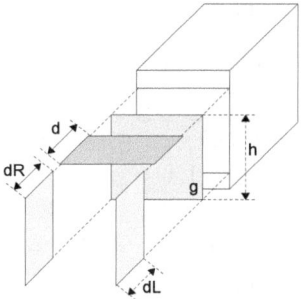

Fig. 1. Graphical representation of variables involved in the optimization problem

be measured under a large set of conditions and constraints. In this research we focus on the following variables, regarded in the literature as the most relevant for the problem: the proportion of glaze in the surface, the depth of different overhangs for shading the windows and the type of glass used. All the considered variables are continuos quantitative variables except for type of glass, which is a nominal qualitative variable. We formulate the optimization problem as the minimization of the energy consumption necessary for maintaining the comfort conditions of a room, regarded as a module of a building. We address this problem by developing a simulation study. We compute the energy loads for heating Q_H, lighting Q_{light} and cooling Q_{cool} adopting the software EnergyPlus[1].

According to Kragh and Simonella [15], we can estimate the amount of carbon dioxide emissions with the following expression:

$$E_{CO_2} = \frac{f_{gas}}{\eta_H} Q_H + f_{el} \left(Q_{light} + \frac{Q_{cool}}{COP} \right) \quad kgCO_2 \qquad (4.1)$$

where Q_H, Q_{light}, Q_{cool} are measured in kWh, $f_{gas} = 0.194\,kgCO_2/kWh$ and $f_{el} = 0.422\,kgCO_2/kWh$ represent the amount of carbon dioxide for the production of 1kWh using natural gas or electricity respectively, $\eta_H = 0.89$ is a measure of the efficiency of the heating system and COP is the performance's coefficient, that is the ratio of useful output to energy input. Here we refer to the London climatic conditions, so we assume $COP = 3.4$ [13].

The variables involved in the minimization problem are:

– h: height, in meters, of the glazed surface, discretized as follows

$$h \in \{0.75 + 0.05i, \ i = 0, \ldots, N_h\}$$

$$N_h = \frac{h_{max} - h_{min}}{h_{step}}, \quad h_{max} = 2.85, \quad h_{min} = 0.75, \quad h_{step} = 0.05$$

---

[1] http://apps1.eere.energy.gov/buildings/energyplus/

Table 1. Parameters of the simulation

Algorithms	Parameters for the quantitative variables	Parameters for the qualitative variable
PSOCC	$\chi = 0.729$, $c_1 = c_2 = 2.05$	3 bit, Gray code, $V_{max} = 4.0$
PSOIW	$0.4 \leq w \leq 0.9$, $c_1 = c_2 = 1.49$	
PSOIWNL	$0 \leq w \leq 0.729$, $c_1 = c_2 = 1.49$	
Q-PSOCC	$\chi = 0.729$, $c_1 = c_2 = 2.05$	$\pi^{gbest} = 0.35$, $\pi^{pbest} = 0.25$, $\pi^{res} = 0.40$
Q-PSOIW	$0.4 \leq w \leq 0.9$, $c_1 = c_2 = 1.49$	
Q-PSOIWNL	$0 \leq w \leq 0.729$, $c_1 = c_2 = 1.49$	

- d: depth, in meters, of the horizontal overhang, discretized as follows

$$d \in \{0.10 + 0.10i, \ i = 0, \ldots, N_d\}$$

$$N_d = \frac{d_{max} - d_{min}}{d_{step}}, \quad d_{max} = 1.00, \quad h_{min} = 0.10, \quad h_{step} = 0.10$$

- dL: depth, in meters, of the left vertical overhang, discretized as follows

$$d_L \in \{0.10 + 0.10i, \ i = 0, \ldots, N_{d_L}\}$$

$$N_{d_L} = \frac{dL_{max} - dL_{min}}{dL_{step}}, \quad dL_{max} = 1.00, \quad dL_{min} = 0.10, \quad dL_{step} = 0.10$$

- dR: depth, in meters, of the right vertical overhang, discretized as follows

$$d_L \in \{0.10 + 0.10i, \ i = 0, \ldots, N_{d_L}\}$$

$$N_{d_R} = \frac{dR_{max} - dR_{min}}{dR_{step}}, \quad dR_{max} = 1.00, \quad dR_{min} = 0.10, \quad dR_{step} = 0.10$$

- g: type of glass, with the following labels

$$g = \{g_0, g_1, g_2, g_3, g_4\}$$

4.1   The Optimization Problem

Without loss of generality, we restrict the domains of variables h, d, dL and dR to a finite set of possible levels in order to achieve reasonable solutions. The variable g, coded by labels g_i, $i = 1 \ldots 4$, is considered as a nominal qualitative variable as it describes specific types of glass. A schematic graphical representation of the variables is reported in Fig. 1.

For designing a building façades we derive Q-PSO introduced in Sect. 3, which allows the consideration of both quantitative variables (h, d, dL and dR) and a qualitative variable (g). Since the search space consists of the finite set of points defined by the Cartesian product of the domains of each variable, the number of possible experimental points is equal to 215 000.

This number of candidate points makes extremely difficult to compute the objective function defined in 4.1 for the whole space as the calculation of Q_H, Q_{light} and Q_{cool} by means of EnergyPlus would involve a very large amount of computational time. We address this problem by considering two different settings, which differ for the number of variables and complexity:

- *Case 1*: we characterize the problem with variables h, d and g, which requires to test 2 150 experimental points;
- *Case 2*: we characterize the problem with variables h, d, g, dL and dR, which requires to test 215 000 experimental points.

The relative small number of experimental points of *case 1* allows the evaluation of all the search space and the identification of the actual minimum value. The structure of *case 1* can then be used as a "test-bed" to evaluate the performance of Q-PSO.

Due to the computational time needed to evaluate the whole search space, for *case 2* we will consider the minimum value that has been found in the simulation study as the problem optimum value.

We address the optimization problem in the following way:

1. Initialization of a random population of P particles, where each particle represents a specific variable configuration for the façade (i.e. an experimental point);
2. Computation of the energy consumption associated to each particle by means of the simulation software EnergyPlus;
3. Computation of the amount of carbon dioxide produced by each particle using formula (4.1);
4. Update the particle positions according to the achieved result;
5. Repeat steps 2–4 T times, where T is a parameter (called the number of generations) fixed by the investigator.

In this simulation we set different values of parameters for the experimentation: for *case 1* we use $P = 30$ and $T = 10$, while for *case 2* we assume $P = 50$ and $T = 30$. With respect to the orientation of the façade, we assume in this setting the East orientation.

Due to the stochastic nature of the optimization techniques, we decide to test the algorithm on 5 different runs in order to validate the results and to evaluate the performance of the optimization procedure. The implementation of the algorithm is realized by using the free software R-project[2].

4.2 The Performance of Q-PSO Optimization Approach

In addressing the problem of designing energy-efficient façades we build Q-PSO with sampling probabilities based on the knowledge achieved on the process and some preliminary tests. In particular, we set:

---

[2] http://www.r-project.org

$$\pi_q^{gbest}(t) = 0.35 \qquad {}_i\pi_q^{pbest}(t) = 0.25 \qquad \pi_q^l(t) = \frac{0.40}{3}, \qquad (4.2)$$

$$l \notin \{L_q^{gbest}, {}_iL_q^{pbest}\}, \ \forall t \in \{1, 2, \ldots, T\}, \ \forall i \in \{1, 2, \ldots, P\} \qquad (4.3)$$

To evaluate the performance of Q-PSO and develop comparisons with other approaches, we implement *binary PSO* as proposed by [3], currently used to deal with qualitative variables. We derive binary PSO encoding the labels of the variable g with a 3 bit Gray code. The comparison between binary PSO and Q-PSO is realized under different hypotheses on the *inertia* weight.

In particular, we consider:

- PSOCC and Q-PSOCC, a constriction coefficient χ is introduced in according with [5];
- PSOIW and Q-PSOIW, linearly decreasing *inertia* weight is considered [8];
- PSOIWNL and Q-PSOIWNL, non-linearly decreasing *inertia* weight is considered [9].

In Table 1 we summarize the parameters, χ, w, c_1 and c_2, of the implemented algorithms for different weight structures. First we consider *case 1*, in order to evaluate the performance of the algorithms when the optimal solution is known, and later we will consider *case 2*, where the complexity of the problem imposes to regard the optimum as 175.60 (the best minimum value obtained in the simulations).

For case 1 we evaluate by simulation all the 2150 possible different façade configurations and achieve the minimum value 179.55, regarded as the target of the optimization procedures. The comparison of the algorithms performance is then derived by computing the average of the minimum values and the minimum value of the objective function determined on the best positions $gbest(t)$ at each generation $t \in \{1, 2, \ldots, T\}$ by each method over 5 runs. The main results are presented in Table 2. We observe that Q-PSO finds better averages of the minimum values with respect to standard PSO. Both the approaches find the minimum value identified as target, but Q-PSO improves the average minimum solution in 5 runs reducing its variability. Q-PSO always reaches the target of the minimization problem, exhibiting a particular stability of the approach.

Table 2. Performance of the algorithms for case 1

	PSOCC	Q-PSOCC	PSOIW	Q-PSOIW	PSOIWNL	Q-PSOIWNL
Target	179.55	179.55	179.55	179.55	179.55	179.55
Average of the minimum value in 5 runs	180.04	**179.57**	180.04	**180.03**	180.48	**179.55**
(sd)	(1.05)	(0.04)	(1.05)	(1.03)	(1.27)	(0.00)
Num. of runs achieving the minimum	3	**4**	3	3	3	**5**

Table 3. Performance of the algorithms for case 2

	PSOCC	Q-PSOCC	PSOIW	Q-PSOIW	PSOIWNL	Q-PSOIWNL
Minimum value	175.60	175.60	175.60	175.60	175.60	175.60
Average of the minimum value in 5 runs	176.05	**175.90**	176.64	**175.89**	175.90	**175.60**
(sd)	(0.45)	(0.66)	(0.61)	(0.64)	(0.66)	(0.00)
Num. of runs achieving the minimum	1	4	1	4	4	5

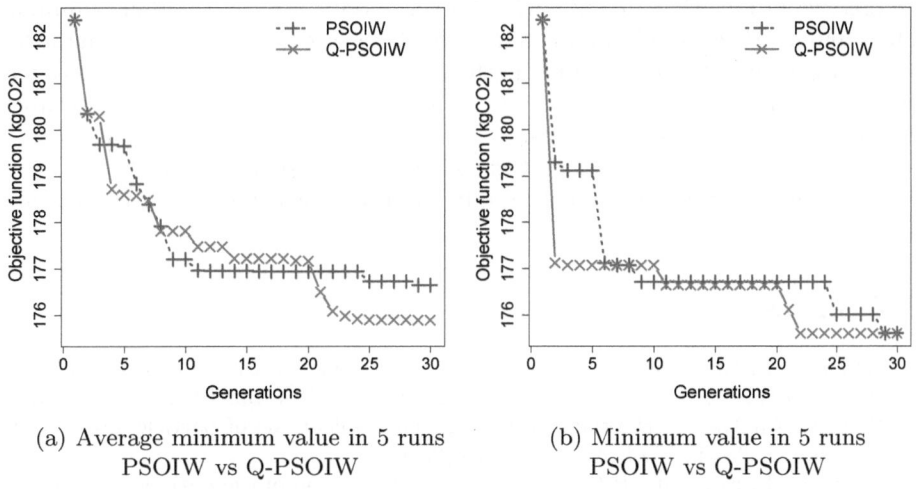

(a) Average minimum value in 5 runs
PSOIW vs Q-PSOIW

(b) Minimum value in 5 runs
PSOIW vs Q-PSOIW

Fig. 2. *Case 2*: comparison of PSO and Q-PSO.

Similarly for *case 2* we should evaluate by simulation the 215000 possible different façade configurations to discover the minimum value of the objective function and regard this value as the target of the optimization procedures. Given the computational time needed to evaluate the objective function we proceed by comparing the behavior of the two approaches in 30 generations and different *inertia* weights. Results are summarized in Table 3. In Fig. 2 we can observe that also for complex problems, as the one proposed in *case 2*, Q-PSO outperforms PSO both in the behavior of the average of the minimum values and in the minimum values, for all the *inertia* weight structures here considered. Moreover, Q-PSO achieves the minimum value in fewer generations than PSO.

5   Concluding Remarks

In this paper we addressed an optimization problem which involves both quantitative and qualitative variables adopting the successful and well-known procedure of PSO. Given the difficulties of this procedure to deal with qualitative variables we introduced a generalization of the procedure: Qualitative Particle Swarm Optimization (Q-PSO), which is a stochastic optimization approach able to deal both with qualitative and quantitative variables. Q-PSO approach is based on the idea of sampling labels of qualitative variables with a probability distribution depending on the global and individual best labels achieved in each generation of the algorithm.

We implemented Q-PSO to the case study of deriving the optimal design of a building envelope able to minimize the carbon emissions due to heating, cooling and lighting, and compared this approach to standard PSO (binary PSO). The study, which involves different structures of the problem, shows that Q-PSO outperforms PSO in almost all the parameter configurations that we considered. We observe in fact that Q-PSO achieves very good results in finding minimum values and averages of minimum values over a set of runs and in few generations of the algorithm. In conclusion, in this study we introduced a new, easy to implement and effective approach that allows considering qualitative variables in complex optimization problems.

As future works we intend to test Q-PSO to different case studies in order to generalize its properties and to study the robustness increasing the number of runs.

References

1. Kennedy, J., Eberhart, R.: Particle swarm optimization. In: Proceedings of IEEE International Conference on Neural Networks (ICNN) 4, pp. 1942–1948 (1995)
2. Poli, R.: Analysis of the publications on the applications of particle swarm optimisation. J. Artifi. Evol. Appl. **4**, 1–10 (2010)
3. Kennedy, J., Eberhart, R.C.: A discrete binary version of the particle swarm algorithm. In: Proceedings of IEEE International Conference on Systems, Man, and Cybernetics 5, pp. 4104–4108 (1997)
4. Laskari, E.C., Parsopoulos, K.E., Vrahatis, M.N.: Particle swarm optimization for integer programming. In: Proceedings of the IEEE 2002 Congress on Evolutionary Computation 2, pp. 1582–1587 (2002)
5. Clerc, M., Kennedy, J.: The particle swarm - explosion, stability, and convergence in a multidimensional complex space. IEEE Trans. Evol. Comput. **6**, 58–73 (2002)
6. Chen, W.N., Zhang, J.: A novel set-based particle swarm optimization method for discrete optimization problem. IEEE Trans. Evol. Comput. **14**(2), 278–300 (2010)
7. Shi, Y., Eberhart, R.: A modied particle swarm optimizer, In: Proceedings of Evolutionary Computation, p. 6973 (1998)
8. Shi, Y., Eberhart, R.C.: Empirical study of particle swarm optimization. In: Proceedings of the 1999 Congress on Evolutionary Computation 3 (1999)
9. Chatterrjee, A., Siarry, P.: Nonlinear inertia weight variation for dynamic adaptation in particle swarm optimization. Comput. Oper. Res. **33**(3), 859–871 (2006)
10. Liao, C.J., Tseng, C.T., Luarn, P.: A discrete version of particle swarm optimization for flowshop scheduling problems. Comput. Oper. Res. **34**(10), 3099–3111 (2007)
11. Rezaee, A.J., Jasni, J.: Particle swarm optimisation for discrete optimisation problems: a review. Artifi. Intell. Rev. 1–16 (2012). doi:10.1007/s10462-012-9373-8
12. Rapone, G., Saro, O.: Optimisation of curtain wall façades for office buildings by means of PSO algorithm. Energ. Build. **45**, 189–196 (2012)

13. Zemella, G., De March, D., Borrotti, M., Poli, I.: Optimised design of energy-efficient building façades via evolutionary neural networks. Energ. Build. **43**(12), 3297–3302 (2011)
14. Ihm, P., Krarti, M.: Design optimization of energy-efficient residential buildings in Tunisia. Build. Environ. **58**, 81–90 (2012)
15. Kragh, M., Simonella, A.: The missing correlation between thermal insulation and energy performance of office buildings. In: Proceedings of the International Conference on Building Envelope Systems and Technology (2007)

Introducing Interactive Evolutionary Computation in Data Clustering

Anna Russo[1], Onofrio Gigliotta[1]([✉]),
Francesco Palumbo[1], and Orazio Miglino[1,2]

[1] Natural and Artificial Cognition Laboratory, University of Naples Federico II,
Naples, Italy
onofrio.gigliotta@unina.it
[2] Institute of Cognitive Sciences and Technologies, CNR, Rome, Italy

Abstract. Data clustering consists in finding homogeneous groups in
a dataset. The importance attributed to cluster analysis is related to
its fundamental role in many knowledge fields. Often data clustering
techniques are the *ghost host* of many innovative applications for a wide
range of problems (i.e. biology, marketing, customers segmentation, *intelligent machines*, machine translation, etc.). Recently, there is an emerging interest in *Data Clustering* community to develop bio-inspired algorithms in order to find new methods for clustering. It is widely observed
that bio-inspired algorithms and the Evolutionary Computation (EC)
techniques reach solutions similar to others computational approaches
but using a bigger computational power. This limitation represents a
concrete obstacle to an extensive use of Evolutionary (or bio-inspired)
approach to data clustering applications. In the present paper we propose
to use Interactive Evolutionary Computation (IEC) techniques where a
human being (the breeder) selects *Cluster configurations* (genotypes) on
the basis of their graphical visualizations (phenotypes). We describe a
first version of a software, called Revok, that implements the IEC basic
principles applied to data clustering. In the *conclusion* section we outline
the necessary steps to reach a mature IEC tool for data clustering.

Keywords: Interactive Evolutionary Computation · Data mining

1 Introduction

Data clustering consists in finding homogeneous groups in a dataset. A very
general Data Clustering definition is the following: given a set of n points in a p
dimensional space, cluster analysis aims at grouping data into k groups that are
maximally homogeneous within each one and maximally heterogeneous between
them [6]. The importance attributed to cluster analysis is related to its fundamental role in many knowledge fields. In fact, on natural side, human and animal
learning/adapting processes produce clusters of objects and/or actions that we
refer to as concepts, mental categories, perceptual patterns, behavioural strategies and so on. On the other hand, data clustering techniques are the *ghost host*

© Springer International Publishing Switzerland 2014
C. Pizzuti and G. Spezzano (Eds.): WIVACE 2014, CCIS 445, pp. 26–36, 2014.
DOI: 10.1007/978-3-319-12745-3_3

of many innovative applications for a wide range of problems (i.e. genetics, marketing, customer segmentation, *intelligent* machines, etc.). Because of their basic importance in different fields, methods for data clustering have been independently developed in very different scientific contexts where a variety of methods have been proposed [6]. Recently, there is an emerging interest in Data Clustering community to build-up algorithms using bio-inspired techniques (Neural Networks, Genetic Algorithm, Evolutionary Computation, Ant algorithms, Swarm algorithms, etc.) [5,17,20,21]. More specifically, there is a consolidated scientific literature concerning EC techniques applied to data clustering [1]. In general, bio-inspired techniques, as all clustering methods, measure and optimize the efficiency of a given algorithm on the basis of an evaluation function. In the case of EC techniques it corresponds to the fitness function. The novelty of EC techniques, respect to more traditional approaches, is represented by the possibility to have a parallel competition between many different clustering variants. The fittest variants are selected for reproduction. The selected individuals generate new variants ready for a new evaluation (or evolution) phase. The evaluation, selection and reproduction processes could be iterated until an efficient solution emerges. It is widely observed that EC techniques produce results similar to other approaches but they require bigger computational resources. Perhaps, all computational approaches (either bio-inspired and traditional) have reached ceiling performances that cannot be overcome anymore. If we adopt a natural cognition point of view this limitation could have an explanation that leads to possible new approaches in Data Clustering. In psychological terms multidimensional data clustering is a kind of categorization based on the analysis of explicit information and quantitative variables (dimensions) that describe a given phenomenon. On the contrary, the best clustering systems for multidimensional domains actually known, the human beings, do not work only on the basis of explicit information. They use also (or mainly) latent and implicit information captured by (neuro)cognitive mechanisms that work under of consciousness threshold. Psychological literature refer to these categorization mechanisms (i.e. sensory analysis, perception, mental reasoning, etc.) as Cognitive Unconscious [9]. According to these research findings all cognitive representations emerge from a dynamic interplay between unconsciousness and consciousness processes. Moreover, it is showed that Cognitive Unconscious mechanisms and mental schemas are very hard to report at conscious level. The interplay of conscious and unconscious of our neurocognitive structures is particular evident in the case of human experts of some specific domain. The human experts are usually trained for many years to recognize (categorize) natural phenomena on both explicit and latent information. For example, take into consideration the classification abilities of a doctor that make a diagnosis. In this condition a doctor reaches his decision (diagnosis) integrating explicit biological indexes captured by technological tools with implicit knowledge that is produced by his/her professional experiences and training. In other words, humans are extremely able to apply sophisticated clustering algorithms but they are not able to explicit them. This gap between explicit and implicit level of cognition could explain why the categorization performance of artificial systems are poorer compared to those performed by human beings. In this work, we propose to improve EC clustering techniques introducing some qualitative evaluation in selection phase. In concrete terms we propose to apply the paradigm of Interactive Evolutionary Computation (IEC) [14,18] to multidimensional data analysis domains. IEC is a well-know technique used to perform optimization through the selection process of a human evaluator. In the framework of data clustering, a genotype describes a clustering configuration (a data partition) and/or its parameters (number of clusters, clusters size, clusters centroids, etc.); the environment is a dataset extracted from a data universe for a given problem; the phenotypic traits are quantitative (i.e.: performances indexes) and/or qualitative (i.e.: pictorial data aggregations) representations of organisms outputs. Individuals are selected for reproduction taking in account phenotypic traits. Following the metaphors, we have now three possibilities:

1. selecting the organisms on the basis of quantitative traits
2. selecting the organisms on the basis of qualitative traits
3. selecting the organism combining qualitative and quantitative traits

To our knowledge, the first option is almost the only method used in Evolutionary Computation techniques applied to data clustering. In concrete terms, a given performance criteria or fitness function is used to automatically select the organisms and the EC techniques are used to minimize or maximize that fitness measure. The other two selection procedures are (to our knowledge) completely new in the field of data clustering while are currently used in other EC applications. These selection techniques require the intervention of a human operator that interacts with the Artificial Evolution process. In other words, a *Breeder* analyses the qualitative phenotypic traits and decides which *organism* is ready for the reproduction. This procedure is usually called Interactive Evolutionary Computation (for a review see [4]; for a recent application to robotics see [13]). To our knowledge there have been only two attempts to introduce IEC in clustering processes. In the first, IEC has been used in order to retrieve images from a database according to a set of features extracted by a set of images presented and evaluated by a human evaluator [10]. In the second paper, IEC has been used in exploratory cluster analysis by showing 2d visualization maps of a set of SOMs [19]. In both works the proposed solutions were suited to specific problems while in the present paper we introduce a general solution feasible for a wide range of clustering problems.

2   Clustering and GA

Clustering is an unsupervised classification technique whose objective is to partition a set of data points into groups or clusters. Resulting clusters should show homogeneity within groups and heterogeneity between groups so that the distance (or any similarity measure) between data points belonging to different clusters be greater than the distance between data points belonging to the same cluster [8]. The most known clustering algorithm is the K-means, which aim is to minimize the within-cluster sum of squares. Two problems affect the K-means algorithm: the first is related to the number of clusters, a choice that a user has to make a priori. The second problem refers to the sensibility to the initial conditions. Different initial conditions led to different partitions by stopping the algorithm in local optima. In order to overcome such kind of problems the use of Genetic Algorithms (GA) has been proposed for clustering data. In particular, GKA (Genetic K-means algorithm), FGKA (Fast GKA) and IGKA have proven to be able to find a global optimal partition given a prefixed number of clusters [17] while Bandyopadhyay and Maulik [2] have used GA to discover automatically also the number of clusters. Clustering, as stated above, is an unsupervised technique that partitions data in order to minimize variance within groups and maximizing variance between groups. This statistical principle is well grounded on the information contained in the data but completely ungrounded on the needs and on the perceptual ability of who have to make use of such partition. In order to solve this issue, in this paper, we introduce an interactive GA clustering. The use of evolutionary search along with the user guide, especially during an exploration phase, can lead to the discovery of interesting and meaningful solutions [14].

3   Interactive GA

In this paper we present an interactive evolutionary algorithm to clustering datasets according to users' needs and perceptual skills. In particular, it permits a user to explore the search space in order to find a suitable partition without specifying the number of clusters. During the initialization phase, a random population of n partitions is created, where n is determined by the users. Each partition is encoded in a genotype string and is displayed to the user through its phenotype: a graphical representation of the found clusters. For each iteration the user is asked to choose its favourite g partitions. Then the g chosen genotypes undergo to a reproduction and mutation process in order to recreate a new whole population. The evolutionary process ends when the user is satisfied by a specific partition.

3.1   Genotype

Each genotype encodes a list of c centroids of $d+1$ dimensions (genes), where d is the number of attributes of a given dataset while the further binary dimension is used to encode the centroid activation. In order to facilitate the clustering process, data points within the dataset are ranged between [0,1]. Only one genetic operator is applied to the genotype: mutation. Mutation is achieved by adding, with a probability p, a value extracted from a normal distribution to each gene. For each genotype then, a partition is created by assigning a data item to the closest centroid.

Individual

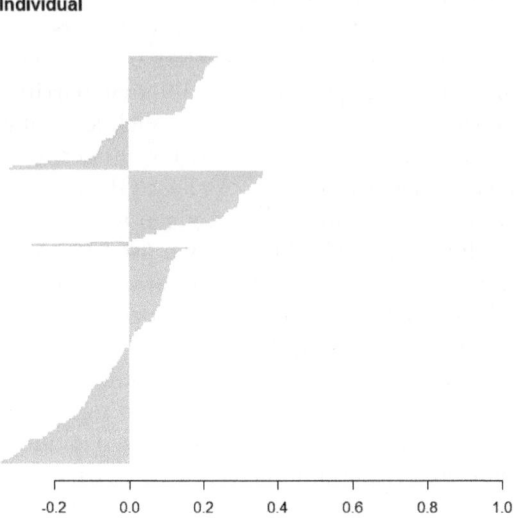

(a) Rousseuw Silhouette

Individual

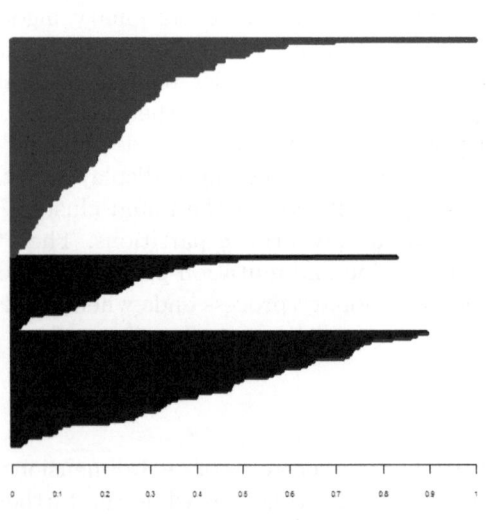

(b) Probabilistic Silhouette

Fig. 1. Silhouette graphs of the same individual (three clusters). In (a) the Rousseuw method in (b) the probabilistic method.

3.2   Phenotype

The genotype-phenotype mapping represents a crucial factor because the phenotype has to describe a solution as clear as possible to the users, even those that are not familiar with sophisticated statistical knowledge, moreover a suitable graphical representation has to allow a user to exploit his perceptual skills in order to discover relations and clusters. For this reasons we decided to use the Silhouette method to graphically represents a partition expressed by a genotype. Originally described by Rousseeuw [16], the Silhouette, computed on distance measures, provides a compact information about how well data items fit into their assigned clusters. Silhouette comes in two formats: based on distance measures [16] and based on a posteriori probability [12]. In the first case, each data item is represented as an horizontal line whose maximum length is 1. A positive length indicates that the item has been correctly clustered while a negative length indicates that the data item is in the wrong cluster (see Fig. 1a). In the probability based Silhouette there are non negative values. This features produces more linear and simple graph (see Fig. 1b). Through a pilot test we observed that the probabilistic approach is more clear in the moment of choice. The principal reasons are:

- the plotted values are positive (a posteriori probability of membership to a specific cluster), in this way the cluster separation is more definite and the user choice is more rapid as we see in Fig. 1. The negative part of classic Silhouette graph is misleading to recognize the number of classes.
- the concept of the probability of membership to a cluster is more intuitive compared to the average of the distances of the classical approach, therefore it is more understandable even in a non-expert user.

4   Revok: Simple Application for IEC in Data Clustering

The IEC data clustering has been implemented in Revok, an application developed in R [11][1] (an interpreted language suitable for statistical purposes) for clustering data through the feedback of a human evaluator. Revok allow users to load their datasets and to set IEC parameters such as, for example, the number of displayable individuals/partitions and mutation rate.

Once started, Revok presents users a series of graphical representations of the partitions found according to the GA. The graphical representation we have chosen for its mathematical properties and its understandable form (humans are very good in perceptual categorization) is the Silhouette calculated according to the method described in [12] by using a posteriori probability computed according to the following equation [3]:

---

[1] The R environment can be downloaded from the internet along with its complete documentation for free. It is an integrated set of software assets for the manipulation and visualization of data. It allows object oriented programming which is particularly suitable for iterative algorithms.

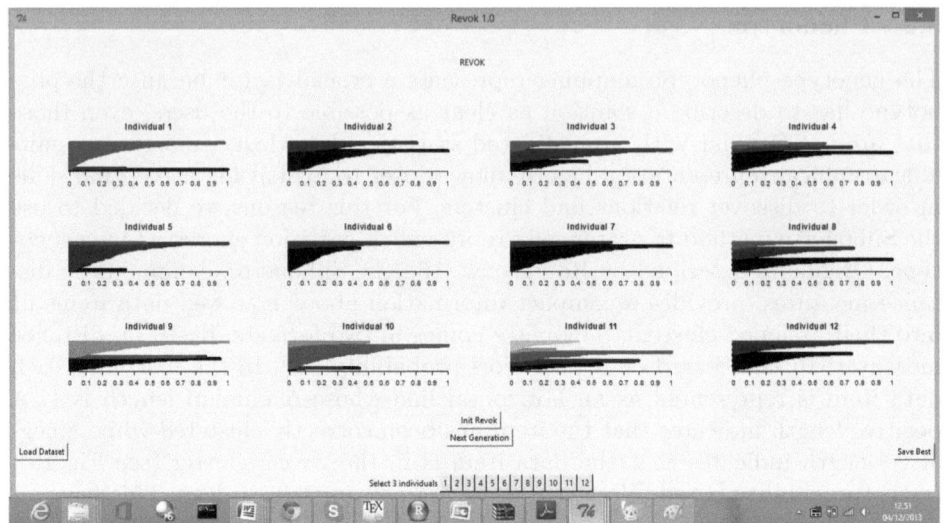

Fig. 2. Display window

$$P_k(x) = \frac{\prod_{j \neq k} d_j(x)}{\sum_{t=1}^{K} \prod_{j \neq t} d_j(x)} \tag{1}$$

Where $d_j(x)$ is the distance of the data item x from the j_{th} centroid and K is the number of centroids. The computed Silhouette is then presented to the user in the display window (see Fig. 2).

In our simple evolutionary algorithm, as previously mentioned, each individual corresponds to a single partition which is the result of a different arrangement of centroids in the solutions space.

The individuals of the next generation are generated from those chosen by the user through a process of mutation. Mutation is applied by adding to each gene an independent random Gaussian number $N(0, \sigma)$, where σ is defined by the user. Revok has been designed to allow users, without any particular mathematical skill, to carry out their own clusterization relying only on their perceptual skills. For this reasons we tested Revok with undergraduate students in psychology, asking them to partition two UCI datasets, Iris and Haberman, following only the request to obtain graphs with a low number of clusters and long horizontal lines. Making use of the mentioned datasets allow us to compare revok users' partitions with those provided with the UCI datasets.

5   Results

Revok has been tested with two datasets: *Iris and Haberman* downloaded from the UCI machine learning repository (http://archive.ics.uci.edu/ml/). In particular we asked to 10 undergraduate students in psychology to partition the two

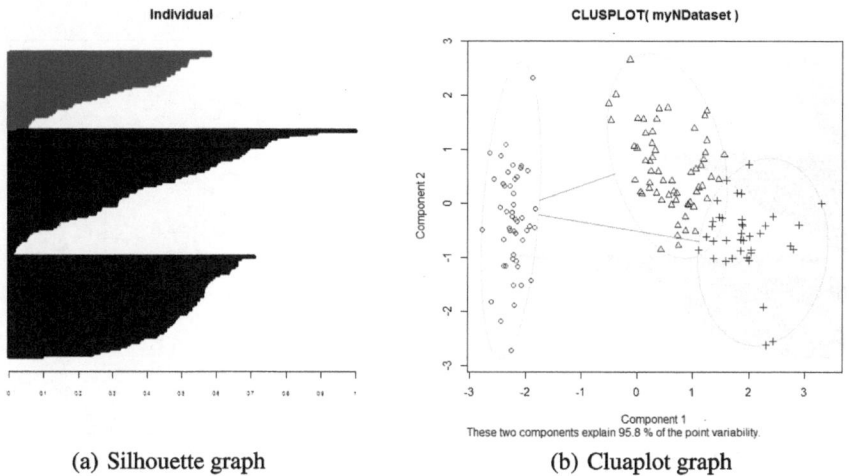

(a) Silhouette graph   (b) Cluaplot graph

Fig. 3. The Silhouette and Clusplot graph for the Iris dataset after five generations with Revok

datasets following only an aesthetic criterion. The following plotted results refer to the typical partitions achieved by those students.

The characteristics of Iris dataset are:

- Multivariate and real dataset
- Number of instance: 150
- Number of attribute: 4
- Number of clusters: 3

The Fig. 3 is a typical result after five generations with the following parameters:

- maximum number of clusters: 6
- mutation rate: 0.2
- σ: 0.1
- number of parents: 1

The figure reports the silhouette and the clusplot graph. The clusplot graph creates a bivariate visualization of the clustered data [15]. All observations are represented by points in the plot, using principal components (PCA). Clusters, then are represented by the ellipses. To represents the distances between the clusters, the clusplot method permits to draw segments of the lines between the cluster centers as reported in our graphs.

The characteristics of Haberman dataset are:

- Multivariate and integer dataset
- Number of instance: 306
- Number of attribute: 3
- Number of clusters: 2

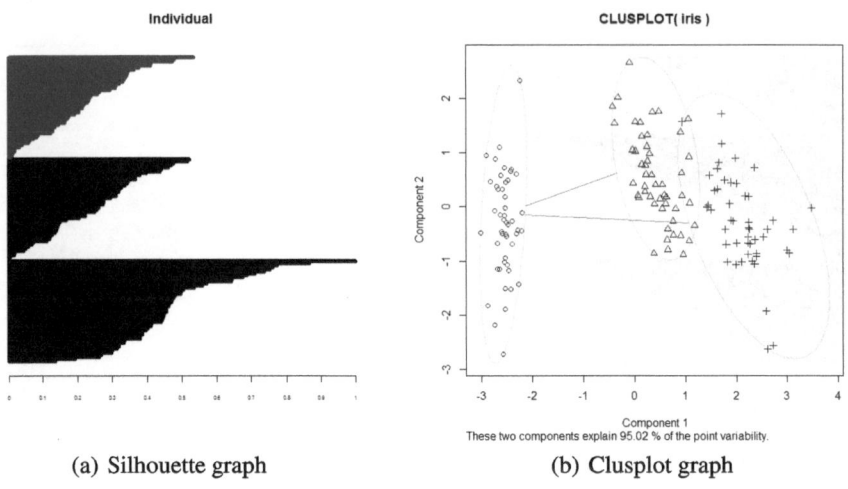

(a) Silhouette graph   (b) Clusplot graph

Fig. 4. The reference Silhouette and Clusplot graph for the Iris dataset

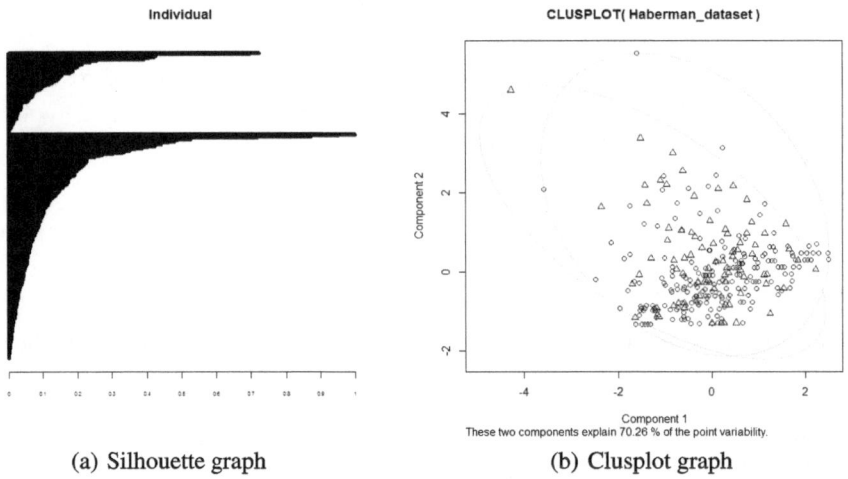

(a) Silhouette graph   (b) Clusplot graph

Fig. 5. The reference Silhouette and Clusplot graph for the Haberman dataset

The Fig. 6 reports a typical result after five generations with the following parameters:

- maximum number of clusters: 6
- mutation rate: 0.2
- σ: 0.1
- number of parents: 1

In the Figs. 4 and 5 are shown the reference silhouette and clusplot graphs for the Iris and Haberman dataset.

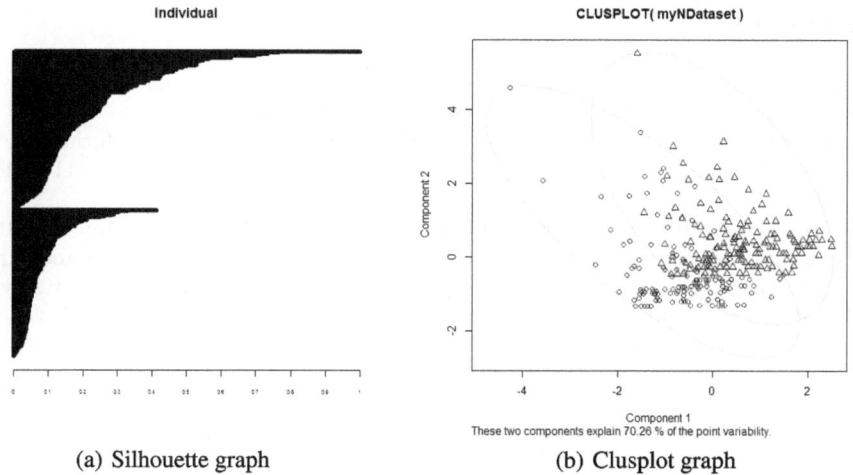

(a) Silhouette graph   (b) Clusplot graph

Fig. 6. The Silhouette and Clusplot graph for the Haberman dataset after five generations with Revok

Results after few iterations, as can be seen from plotted graphs, approximate well the real partitions of the two datasets.

6   Conclusions

Revok is still a proof of concept and we need to test it with many more datasets, but clearly shows how IEC, along with compact visualization tools, can help users in clustering data by simply relying on their perceptual skills. Moreover, IEC can help users to ground the clustering process on their needs and knowledge (even if this knowledge is almost unconscious [9]). In fact, when living animals categorize incoming information, they categorize not only by virtue of the statistical property of the data flow but, above all, for their purposive behaviour [7]. The importance of IEC in data clustering resides in this human value. In order to improve Revok thus, we can act at different levels. A first level refers to distance between data points, usually we utilise euclidean distances but studies about the formation of concepts in humans, suggest that we may use different metrics: future version of Revok will be able to capture such a feature. A second level is about the visualization. At the moment each data point is shown as a line in the silhouette graph, but it can be useful to see data items in terms of physical features (i.e. color, shape etc.) in order to facilitate the recognition of similarity between elements.

References

1. Abul Hasan, M.J., Ramakrishnan, S.: A survey: hybrid evolutionary algorithms for cluster analysis. Artif. Intell. Rev. **36**(3), 179–204 (2011)
2. Bandyopadhyay, S., Maulik, U.: Genetic clustering for automatic evolution of clusters and application to image classification. Pattern Recogn. **35**(6), 1197–1208 (2002)
3. Ben-Israel, A., Iyigun, C.: Probabilistic d-clustering. J. Classif. **25**(1), 5–26 (2008)
4. Brintrup, A., Ramsden, J., Tiwari, A.: A review on design optimisation and exploration with interactive evolutionary computation. In: Tiwari, A., Roy, R., Knowles, J., Avineri, E., Dahal, K. (eds.) Applications of Soft Computing. AISC, vol. 36, pp. 111–120. Springer, Heidelberg (2006)
5. Du, K.L.: Clustering: a neural network approach. Neural Netw. **23**(1), 89–107 (2010)
6. Everitt, B., Landau, S., Leese, M.: Cluster Analysis. A Hodder Arnold Publication. Wiley, New York (2001)
7. Goodwin, C.J.: A History of Modern Psychology. Wiley, New York (2002)
8. Hruschka, E.R., Campello, R.J.G.B., Freitas, A.A., De Carvalho, A.C.P.L.F.: A survey of evolutionary algorithms for clustering. Trans. Syst. Man Cybern. Part C **39**(2), 133–155 (2009)
9. Kihlstrom, J.: The cognitive unconscious. Science **237**(4821), 1445–1452 (1987)
10. Lee, J.Y., Cho, S.B.: Sparse fitness evaluation for reducing user burden in interactive genetic algorithm. In: 1999 IEEE International Fuzzy Systems Conference Proceedings, 1999. FUZZ-IEEE '99, vol. 2, pp. 998–1003 (1999)
11. Lumley, T.: R Fundamentals and Programming Techniques. Chapman & Hall/CRC, Boca Raton (2006)
12. Menardi, G.: Density-based silhouette diagnostics for clustering methods. Stat. Comput. **21**(3), 295–308 (2011)
13. Miglino, O., Gigliotta, O., Ponticorvo, M., Stefano, N.: Breedbot: an evolutionary robotics application in digital content. Electron. Libr. **26**(3), 363–373 (2008)
14. Parmee, I., Bonham, C.: Cluster-oriented genetic algorithms to support interactive designer/evolutionary computing systems. In: Proceedings of the 1999 Congress on Evolutionary Computation, 1999. CEC 99, vol. 1, pp. 546–553 (1999)
15. R Development Core Team: R: A Language and Environment for Statistical Computing. R Foundation for Statistical Computing, Vienna, Austria (2008). http://www.R-project.org, ISBN 3-900051-07-0
16. Rousseeuw, P.J.: Silhouettes: a graphical aid to the interpretation and validation of cluster analysis. J. Comput. Appl. Math. **20**, 53–65 (1987)
17. Sheikh, R., Raghuwanshi, M.M., Jaiswal, A.: Genetic algorithm based clustering: a survey. In: First International Conference on Emerging Trends in Engineering and Technology, 2008. ICETET '08, pp. 314–319 (2008)
18. Takagi, H.: Interactive evolutionary computation: fusion of the capabilities of EC optimization and human evaluation. Proc. IEEE **89**(9), 1275–1296 (2001)
19. Teh, C.S., Chen, C.J.: Interactive evolutionary computation and density-based clustering for data analysis. In: International Conference on Intelligent and Advanced Systems, 2007. ICIAS 2007, pp. 104–108 (2007)
20. Xu, R., Wunsch II, D.: Survey of clustering algorithms. Trans. Neural Netw. **16**(3), 645–678 (2005)
21. Yang, Y., Kamel, M.S.: An aggregated clustering approach using multi-ant colonies algorithms. Pattern Recogn. **39**(7), 1278–1289 (2006)

Living Emerging Worlds for Games

Nicolas Jakob$^{(\boxtimes)}$ and Carlos Andrés Peña

School of Business and Engineering Vaud (HEIG-VD), University of Applied Sciences
and Arts Western Switzerland (HES-SO), Yverdon, Switzerland
nicow.jakob@gmail.com

Abstract. In this paper we propose an agent-based, artificial-life inspired, model for living emergent worlds intended to provide realistic
behaviors to, already graphically sophisticated, game environments. The
ultimate goal of such a proposal is to create the impression to play in
a real, living environment. The model is conceived as a physical world
where several layers of geographic characteristics and sources of influences define the environment for the living entities. These former are
agents that react, based on very simple rules, to the state of the environment and to the influences of other agents. The combined behavior of all
these agents produces a dynamic environment that changes gently, but
constantly in an unpredictable, although plausible, manner. The proof
of the concept is established through the simulation of such a world performed on a fully-functional implementation of the model. We present an
example of a simulation and discuss the encouraging results of this and
other simulations. These results show that it is possible to create complex and stable ecosystems using simple sets of rules. However, it appears
also clear that setting some of the parameters to obtain an accept-
able behavior is a delicate task that deserves further exploration and
development.

Keywords: Emerging systems · Dynamic game ecology · Artificial life

1 Introduction

In order to provide realistic environments, modern games rely mainly on graphic
rendering rather than on nature-like ecosystems. Because of that, players face
static virtual worlds that exhibit little, if any, changes as all is predefined by
design. An increasingly used alternative is the one proposed by, so-called, sand-
box games that provide players the ability to affect permanently the world. The
question that arises, and that is addressed by the current work, is: why not simply, or in addition, enabling the world to change by itself? This paper focuses
thus on techniques that model living emerging ecosystems as background environments for games aiming at providing players with dynamic environments —
i.e., systems that change gently but constantly according to the normal development of their living components such as creatures or plants.

© Springer International Publishing Switzerland 2014
C. Pizzuti and G. Spezzano (Eds.): WIVACE 2014, CCIS 445, pp. 37–46, 2014.
DOI: 10.1007/978-3-319-12745-3_4

This paper is organized as follows. The remaining of this section discusses some lack of realism, from an ecological behavior point of view, observable in current landscapes or environments. Then, the next Section presents the main ideas of our proposal intended to improve realism through emerging living systems. Afterwards, it describes several features that should allow building living emerging worlds, enabling game developers to create ecosystems closer to reality than pre-generated worlds. The paper continues presenting and analysing the results of some experiments conducted on the prototypical implementation. Finally, the last section draws some conclusions and sketches possible ways to continue the work.

1.1   Static Environments in Games

In order to completely immerse players into a game, it is necessary to present a game environment as close as possible to reality. Usually, the expected realism is brought through carefully conceived graphic enhancements. However, they are rarely accompanied by the corresponding, realistic behavior of the environment (e.g., natural cycles of growing, life, and death). Some recent products start to timidly bring some realistic behaviour to the game environment, as is the case of Far Cry [1]. Usually, all the decisions about the entities present in a given environment are taken in advance and defined following some strict rules. During the development process, someone decides where these entities will be placed by manually designing logical clusters like forests, mountains, etc. causing a static distribution of landscape. A first attempt to palliate this limitation has been proposed in some games like Minecraft [7], where the designer, or an algorithm, predefines some *biomes*—i.e., logical areas that share some features and/or conditions—in order to automatically generate appropriate flora and fauna. Even if this technique is an important improvement for creating more realistic environments, it still exhibits some limitations as discussed below.

Biome shapes. Most of the time, biomes are defined as static areas, making it difficult to place entities that ensure smooth transitions between different kinds of biome. The consistency of neighbouring areas can also be an issue as it is unrealistic to suddenly jump, for instance, from an ice plain to a jungle.

Frozen ecosystems and persistence of changes. From a player's perspective, it might be frustrating to walk in a region several times and never see some changes made either by themselves or simply occurring naturally. Furthermore, whenever a change is made, it is natural to expect it to become persistent in time. For instance, the creation of a new city or the destruction of a forest are noticeable events that should persist to reinforce the realism of a game.

Development process. Finally, it is difficult for developers to bring new game components without redesigning significantly, if not completely, the game environment so as to keep it consistent.

Note that some games, like Embers of Caerus [2, 3], are working on the development of a dynamic game ecology, a subject closely related to the work proposed herein.

2   Living Emerging Worlds in Brief

Keeping in mind the goal of offering realistic game environments from an ecological behavior point of view, the simplest manner to solve (some of) the main difficulties described previously is to use environmental models that are close enough to real living environments. Besides, rather than specifically defining how the environment is supposed to behave in every area, why not giving to every entity in the environment the ability to behave by itself according to its nature and to its interactions? These two considerations call for solutions inspired on Artificial Life. Note, however, that while pure Artificial-Life approaches tend to explore complex models and/or behaviors, our proposal intends to address an important issue in games respecting some constraints in terms of computational cost and maintainability. Our general idea turns, thus, around an agent-based environment — conceived as a background service for games — where the living behavior of each agent is driven by a simple set of independent rules. These rules define, for example, if a given area is suitable for a living component according to the current environmental state (See Fig. 1). All agents are, however, not necessarily living components; some of them may only have the purpose to bring essential resources for others. As a simple example, imagine a system composed by three kinds of agents: water sources, trees, and rabbits. There is a clear dependency between them because rabbits require vegetation to survive and trees require water to grow. It is possible to build a system that behaves according to these simple rules and let the system find how to organize these agents.

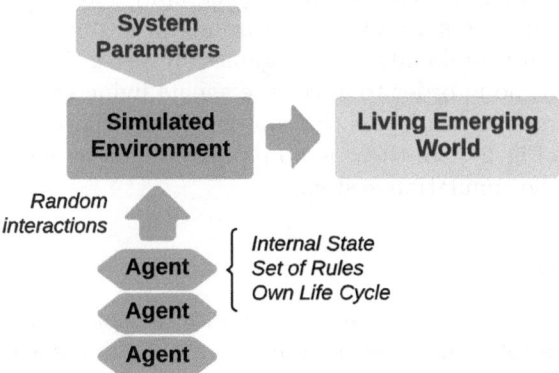

Fig. 1. General idea for emerging living worlds. The global environmental behavior is determined by the cumulated actions and interactions of individual agents. A simple set of rules define how such agents behave according to the influence of their environment.

Describing rule-based behaviors for individual agents is simpler than describing all the possible interactions by hand.

Unpredictability of content. In complex systems, locally-simple behaviours that make sense at an individual level, may bring unexpected results that a human mind might not be able to design manually or to predict precisely. In this context, the term emerging means that complex interactions between agents are brought by the execution of simple behavior blocks.

Introduction of new features. It is possible to easily introduce new rules into a system (i.e., a game) while it is in production without being obliged to take it out of service. It is, however, fundamental to monitor the impact of such changes on the system given that, as mentioned before, the result might be unpredictable.

3   Design Considerations

The proposed concept [5] turns around the interactions between living and environmental agents. So, the design of the sought-after living emerging worlds will depend on the features of both, the environment itself and the agents that populate it.

3.1   Environmental State

The environment, by definition, is a space that must support different kinds of living entities (species). It is expected to be diverse enough so as to contain different biomes occurring naturally without artificial boundaries between them and where living entities are able to live according to their characteristics and behavior rules. To attain this goal, the terrain is defined by means of a set of environmental parameters, determining its geography and its climate. The initial values of these parameters are obtained through procedural content-generation algorithms which are commonly used in games [4] — e.g. Perlin Noise [6]. They may change over time in order to affect the agents living in there. The environmental state is organized in layers that automatically shape areas where agents may appear (See Fig. 2). So, there is no need to define by hand the location of game objects in the simulation system.

3.2   Areas of Effect

The environmental parameters presented above do not permit to model completely all the possible sources of interaction. All the living agents may be able to affect and be affected by every other agent that is close enough. For instance, as illustrated by Fig. 3, a tree will provide vegetation, which can be used by other surrounding agents. Every agent may emit around it a number of influences that permit to create indirect dependency between agents. Indeed, agents react to the emitted influences according to their individual rules — e.g., needs,

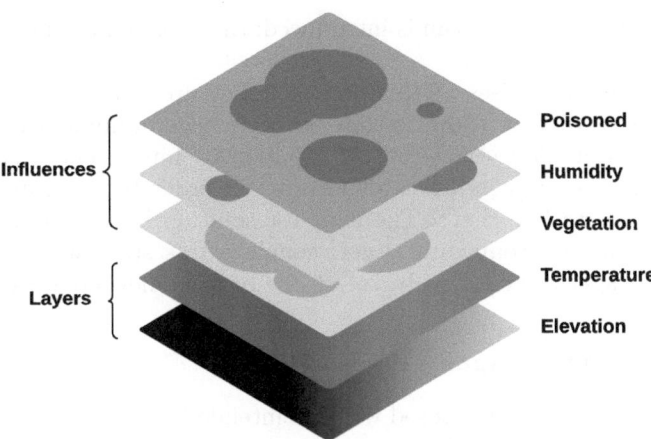

Fig. 2. The state of the world is organized in layers of environmental parameters and influences.

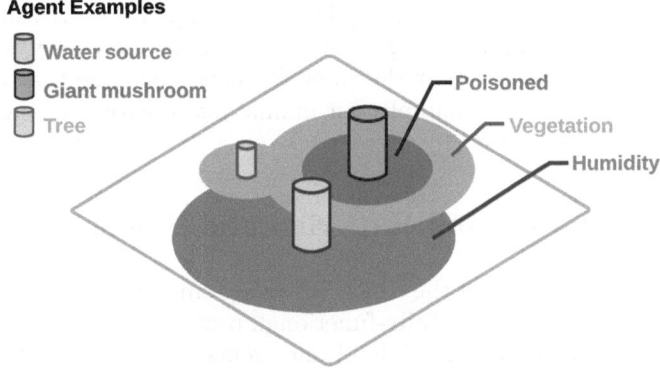

Fig. 3. Agents emit one or more types of influences around them, that may affect other nearby agents. This creates a network of interactions and dependencies among agents.

susceptibilities, weaknesses, or strengths. Moreover, as influences are attached to and managed by each agent, it is possible to use its internal state (age, health, etc.) to change over time the emitted effect. While a plant grows, its vegetation influence will expand too, for instance.

3.3   Live and Death of Agents

As explained before, each individual agent manages its own internal state by sensing its environment. Depending on the environmental state and the surrounding influences, an agent is able to define how suitable is the current area for it to live in. According to the agent's rules, all these influences affect its internal status and may also trigger actions. As agents must not be eternal, an age-related death mechanism is introduced: the probability of dying increases with time.

The question that arises at this point is: How are new agents generated to replace those who die? i.e., how do they born? Instead of implementing complex and costly reproduction strategies, we decided to implement a strategy that periodically checks the conditions of an area in order to generate new agents that are adapted to live there. The simplest solution is to randomly select an area, compute its environmental state, compare this state with the profile of every available species and, finally, select one or more new creatures.

3.4   Life Rhythm of Agents

Keeping in mind that the proposed world is intended to constitute a background environment for games, its simulation should be done in real time as players would be able to influence the environment. Continuously computing the internal state of a large number of living agents in the world is computationally costly and may prevent the expected real-time simulation. To palliate to this, we decided to rationalize the amount of status updates by defining an intrinsic rhythm of life for agents. According to the agent species, it is not necessary to update its internal state (e.g. size, color) at the same frequency because they do not have the same life rhythm. For example, an animal-like creature changes its state much more often than a tree.

4   Implementation and Experimental Results

In order to test the validity of the conceived living emerging world, we developed and implemented a simple, but fully-functional, prototype that allows simulating a relatively large area with many inhabitant agents [5]. The simulator allows the designer to define different species of agents and sources of influences, together with their corresponding set of rules. Note that the simulator is oriented towards managing the world rather than to its graphic rendering. Because of that it provides a simple command line interface to interact with it.

To illustrate the functioning of the prototype simulation environment, let it have a simple ecosystem inhabited by four different kinds of species, as described by Table 1.

The resulting population dynamics is shown by Fig. 4. It illustrates the final distribution of the populations in the world as well as the amount of time, in arbitrary units, required by the ecosystem to begin stabilizing after having started from an uninhabited world. However, in order to attain such stability, it is important that the underlying environmental state does not change too much during this transient phase. Note that, although four species are available (i.e., defined by the designer) only three of them populate the proposed world. This fact derives from the environmental conditions of the world which are not adequate for the cactus species to grow.

This example shows how a simple set of rules allows generating a world populated by different, but interdependent, species adapted to the reigning environmental conditions. As illustrated in Figs. 4 and 5 the agents are not distributed

Table 1. List of species and their characteristics used in the exemplar experimental simulation.

Name	Rhythm	Emissions	Requirements
Tree	Slow	High vegetation	Medium elevation
			High humidity High vegetation
Shrub	Medium	Medium vegetation	Medium humidity
Cactus	Medium	Medium vegetation	High temperature Low humidity
Rabbit	Fast	—	Medium elevation High vegetation

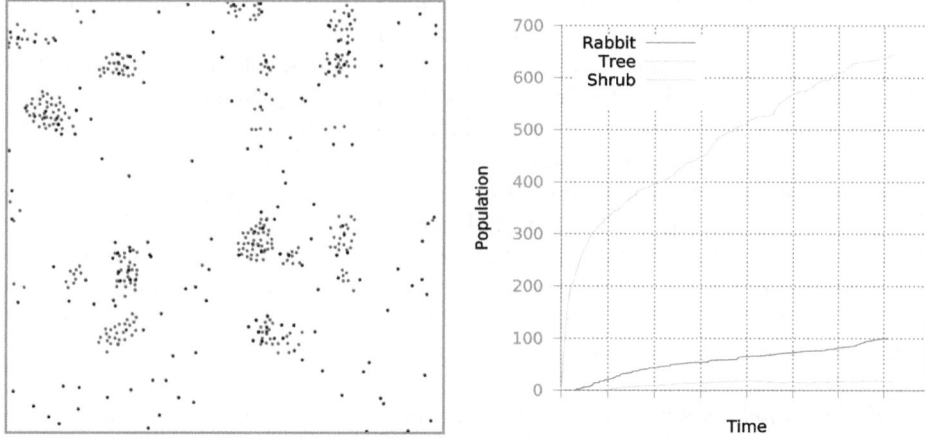

Fig. 4. Example simulation: Three species populate the world during the transient initial phase shown in the figure at the right. At the left, the distribution of the three populations at the end of this simulation period.

Fig. 5. Example simulation: Final distribution of the three populations across the world. From the left: shrubs, rabbits and trees.

arbitrarily through the world, but form small ecological communities. From the figures one can observe four of such niches formed mainly by shrubs and rabbits. Trees are much rarer and they are even absent in one of them.

On the base of this and several other simulations performed under different conditions, we have made the following observations:

Dynamics of the emerging ecosystems. Being a certain stability one of the pursued goals, it is normal that the emerging patterns of agents will generally remain the same until the local environment changes. Therefore, the rapidity at which an ecosystem changes is mainly defined by the rate of change of the environmental parameters. It appears also clear that the final effect of such changes — i.e., the new point of stability — depends greatly on the sensibility of the agents to these parameters.

Population management. We observed that managing the total amount of agents based only on the species' rules, without explicitly defining a limit, is challenging. In many cases the population of a species will simply explode (or implode) as there is no mechanism to reduce the attractiveness of an area when it is already saturated or close to be it. To deal with this problem, we considered two main possible solutions: (1) introducing a producer/consumer system to manage the available resources and (2) setting up a feedback mechanism into the process of agent creation, based on local population densities. Both of these alternatives would imply an additional source of complexity and were, thus, abandoned. Instead, we implemented much lighter variants of them: On the one side, attenuation of the influences as they are "consumed" by affected agents and, on the other side, reduction of the probability of creating new agents based on the distance of the closer existing agents.

Simulation benchmarking. The configuration of a new emerging world as well as the definition of rules for the different species imply conducting a large number of simulations to observe and tune the general behavior of the world. Several resulting aspects should be monitored, and for long simulation periods, in order to assess the adequacy of a given configuration — e.g., resulting populations, geographic distribution, computational cost related with the updating frequency. All this asks for the development of automatic test protocols with quality measurements as an outcome.

5   Concluding Remarks

The results of the performed simulations are encouraging enough as they tend to show that obtaining realistic game environments based on simple rule-based agents is feasible. These simulations have also allowed us identifying some concerns and issues that should be addressed in future developments of our work. The most relevant of them are discussed below.

Emergence versus design. While the approach of using emerging behaviors reinforces the expected non-deterministic realism of the world, it also makes it hard to force specific results. If one wants the system to move towards a precise type of results, a possibility is to introduce dependency between agents, for example in the form of influences.

Setting and adjusting environmental parameters. When configuring a model, many parameters must be set up and/or adjusted simultaneously. However, due to the strong interdependency of the agents among them and with the environment it is difficult to foresee the effects of such adjustments on the whole system behavior. The only manner available to analyze these effects and prevent unacceptable solutions, is to run multiple simulations under controlled conditions. This implies a high computational cost.

System stability and stagnation. Thanks to the live/death cycle of the entities, the whole population is periodically replaced, causing the renewal of the emitted influences, that is beneficial to escape from stagnating conditions. However, such system destabilization should also come from changes in the environmental parameters — e.g., temperature, humidity, elevation. Indeed, the fixity of these sources of influence have a strong effect on the tendency of the simulated world to converge towards the same results.

Integration in games. Knowing that the ultimate goal of our work is to integrate the model in games, it is crucial to define how the players will interact with the living emerging world. On the one side, the world is intended to be dynamic and reactive to create the impression of a living environment. On the other side, the players, according to game requirements, should not be allowed to destabilize excessively the system by (massively) killing creatures or destroying plants, affecting strongly the available influences in a given area. Such event may render entire areas unsuitable for one or more species. This issue should be carefully addressed to ensure that the delicate equilibrium between realism and security is kept.

References

1. Far Cry 3 Team. Open World, Biomes, Vegetation, and Fauna (2012). http://far-cry.ubi.com/fc-portal/en-au/community/detail_news.aspx?c=tcm:22--62197&ct=tcm:6--231-32
2. Forsaken Studios. Embers of Caerus Dev Blog: Dynamic Ecology (2012). http://www.indiedb.com/games/embers-of-caerus/features/dev-blog-dynamic-ecology
3. Forsaken Studios. Embers of Caerus: Dynamic Ecology (2012). https://www.kickstarter.com/projects/forsakenstudios/embers-of-caerus-investor-prototype/posts/248961
4. Hendrikx, M., et al.: Procedural content generation for games. ACM Trans. Multimedia Comput. Commun. Appl. 9(1), 1–22 (2013). http://dl.acm.org/citation.cfm?doid=2422956.2422957, ISSN: 15516857

5. Jakob, N.: Living Emerging Worlds for Games. MA thesis. School of Management and Engineering Vaud, Yverdon-les-Bains, Switzerland, Feb 2014
6. Perlin, K.: Improved Noise Reference Implementation (2002). http://mrl.nyu.edu/~perlin/noise/
7. Persson, M.: Minecraft (2009). https://minecraft.net/game

Studying the Evolutionary Basis of Emotions Through Adaptive Neuroagents: Preliminary Settings and Results

Daniela Pacella[1]($^{\boxtimes}$), Onofrio Gigliotta[2], and Orazio Miglino[2]

[1] School of Computing and Mathematics, Plymouth University, Plymouth, UK
daniela.pacella@plymouth.ac.uk
[2] Natural and Artificial Cognition Laboratory, University of Naples Federico II,
Naples, Italy
{onofrio.gigliotta,orazio.miglino}@unina.it

Abstract. We propose a method to investigate the adaptive and evolutionary function of emotions and affective states, in our case of ancestral fear - using Artificial Life and Evolutionary Robotics techniques. For this purpose, we developed a hybrid software-hardware capable to train artificial neuroagents equipped with a sensory-motor apparatus inspired on the iCub humanoid robot features. We trained populations of these agents throughout a genetic algorithm to perform a well-known neuropsychological task adapted to study emotional phenomena. The robots learnt to discriminate stressful emotional conditions (coping with "dangerous" stimuli) and no-stress conditions. Varying the network structures, the experimental conditions and comparing the outcomes we were able to delineate a very initial snapshot of behavioral and neural prerequisite for emotional-based actions. On the other hand, we have to stress that the main contribution we brought is setting-up a methodology to support future studies on emotions in natural and artificial agents.

Keywords: Artificial Life · Neural networks · Emotion · Neurorobotics

1 Introduction

The importance of studying emotions in animals for a deep comprehension of natural organisms' behavior was denied for a long time. Behaviourists and their experimental paradigms tended to demonstrate that the desired behavior of an animal could be obtained simply through conditioning, regardless of the emotional state experienced. The fear of anthropomorphizing the research on animals also led to a neglect of the evolutionary basis of the emotions, and their function itself was reduced to a cognitive need of labelling autonomic arousals [5]. With the rising of the neuroscientific approach, the attention shifted more to the neural aspects of behavior, and in recent years, thanks to the latest tools designed to investigate the functional organization of the cortical and subcortical structures such as EBS and neuroimaging, the similarity of the arousal mechanisms in humans and animals became more than evident [1,8]. Unfortunately,

© Springer International Publishing Switzerland 2014
C. Pizzuti and G. Spezzano (Eds.): WIVACE 2014, CCIS 445, pp. 47–57, 2014.
DOI: 10.1007/978-3-319-12745-3_5

the problem of studying emotions in animals mainly lies in the difficulty of measuring and detecting variables. The methods used mostly incur into different kind of biases. In the case of the studies on the emotion of fear, for example, it is hard to isolate the correct trigger event which aroused the rat, and the measured variation of responses may be due to other factors, like daily and individual fluctuations. Also, autonomic responses are often similar among totally different kind of emotions: fear and sexual activity share the increase of cortisol level and cardiac frequency. Despite these issues, the research in the field of the neural basis of the emotions went far, and, thanks to the discover of the limbic system and of the central role of the amygdala, LeDoux in his late 90's studies came up with the first model of the conditioned fear circuit in rats [7]. The importance of subcortical structures for the genesis of fear was also proved in humans [3]. Contemporary studies also showed that decorticated rats were still capable of expressing fear [6]. Another well-known problem concerns the definition and operalization of the emotions themselves. Regarding our terminology of fear, affective states and emotion, we refer to Panksepp's nest-layered model of the brain [12,14–17]. It states that we can distinguish three main structures inside the brain with different evolution levels and a hierarchic localization; every region is able to generate a specific set of affective states different in order and complexity from the others. Hence he distinguished three kinds of emotional circuits: *primary processes*, placed subcortically, immediate and independent from cognition (just like conditioned responses), *secondary processes*, which involve the limbic system, and *tertiary processes*, exclusively neocortical and only evidenced in humans, which include the subjective experience of complex emotions the way as we know it [14,17]. The author, pursuing his will to investigate animal emotions, defines 7 types of proven primary processes whose circuits were found in rats: SEEKING, RAGE, LUST, FEAR, CARE, PANIC/GRIEF and PLAY. He uses capitalized letters to distinguish these basic emotional responses from the complex, cortical ones (tertiary processes) of which he has found no evidence in other species than humans [17]. These primary circuits are the proof of the existence of affective states in animals, considered as evolutionary products able to improve the performance, the adaptiveness and the survivability of an agent.

Though Panksepp's and LeDoux's models seem to rely on the totally genetic basis of the emotional responses, Panksepp himself has recently highlighted how important is to define what of an organism is genetic and what epigenetic [13]. For this purpose, he suggests the use of robotics and Artificial Life: using a computation model of the primary processes, the interactions of agents and environment can enlighten us about which role the evolution plays for the genesis of the complete experience of an emotion. Meanwhile, studies involving artificial neural agents mainly focused on reproducing purely cognitive or motivation-driven behaviors as well as at emphasizing the complex structure of the sensory-motor apparatus, while little or no attention was given to the analysis of the models of emotional and affective states, with the exception of few preliminary works. Parisi et al., in fact, recently introduced a new kind of neuron in their networks whose activation threshold differs from the other standard units and able, when active, to directly communicate with the output layer and to gain priority of action, bypassing the hidden units' computation. This immediate information processing which leads to action - is considered of the same kind of the activation of natural organisms' emotion circuits. As we mentioned before, in fact, these mechanisms are effective even in absence of cognition, i.e. without the cortical influence. In their work, they proved that the presence of this particular neuron, which they called emotion unit, significantly improved the survivability and adaptiveness of the neural agents allowing them to efficiently escape from predators or correctly prioritizing between feeding or hiding [18]. The authors of the mentioned preliminary study, however, do not provide a specification of the notion of emotion they refer to, and therefore they do not specify which emotions can be investigated with this paradigm. In our Artificial Life study we inherit the concept of emotion unit to set up the basis of a software able to simulate the evolution of a population of artificial agents; these robots are trained to avoid self-harming stimuli in order to spot the phylogenetic function of genetically evolved fear. We do not focus on the concept of expressing emotions, but our aim is to highlight the contribution that affective states bring to adaptation and survivability of artificial as well as natural agents in a given environment. We therefore refer to the above-mentioned concept of FEAR for setting up our experimental condition: the simulated humanoid agents (based on the iCub robot structure [9, 19]) are subjected to a cancellation task and evolved to avoid a certain type of stimulus if a specific external input is given (the presence of a light) as it becomes dangerous (causes the termination of the robot's lifecycle). Thus, following a distinction made by Canamero [2], our work can be framed into the 'emergent approach' to the study of emotions.

2 Materials and Methods

2.1 The Neuroagent

The iCub humanoid robot is an artificial tool created to reproduce the cognitive and motor aspects of a 3-years old child. His dimensions are realistic: his height is 104 cm and weight is about 22 kg. His body has 53 d.o.f. of which 38 are placed in the upper part. His hands are extremely sophisticated and the head is provided with a visual system which allows him to follow and grasp target objects in his visual field (Fig. 1). The hardware comes up with an open-source development kit which allows to program and to transfer computer simulations on the physical robot to test it in a real environment. In order to perform the task described below, the neuroagents we used are provided with the same sensory-motor features, which grant them the ability to explore the visual scene, to discriminate the type of items presented and to point at selected targets. To simulate our artificial robots and neurocontroller, and to set up the experimental setting, we used a modified version of the software Evorobot*, developed by Nolfi and Gigliotta and described in [11]. This tool allows the user to simulate the behavior of big populations of robots in the specified environment conditions; besides, it is possible to evolve these neuroagents with the use of genetic algorithms, selecting throughout the evolution those with the better fitness and adaptiveness. This selection process can be carried out for hundreds of descending individuals. The platform allows these simulations to be easily transferred on the real robot, placing itself as a hybrid software-hardware tool [10].

The Sensory-Motor Apparatus. The neuroagent's visual system, as described above, allows the detection, identification and reaching of the targets in the experimental condition. In order to perform these actions, every artificial organism is equipped with a pan/tilt camera whose receptive area is a maximum of 350 × 350 pixels. The camera, moving his receptive field horizontally and vertically, allows the robot to detect the stimuli, which are distinguished according to their luminance. Also, the visual apparatus has a zoom, which represents a tool to better focus and analyse the presented stimuli on the visual scene, whose dimension is 400 × 400 pixels. The information perceived by the camera is computed by an artificial retina composed of 49 neurons; each neuron is able to process a visual area of 25 × 25 pixels [4].

The robot is also provided with a motor module which guides the action of selecting the targeted stimulus.

The Neural Controller. The network we used has a typical three-layered structure and every unit has a sigmoid activation function. The input layer is composed by 49 visual neurons and a special computational unit which allows the robot to avoid the dangerous stimuli. This unit, whose function and use have been introduced and described in [18], has an activation function different than the others and, when its threshold is reached, gets the priority over the other actions and triggers the avoiding behavior of the harming type of stimuli. This particular neuron was defined "emotional unit". The neural network hidden layer is provided with 20 hidden units, while the output layer consists of 4 different units: 2 motor neurons controlling the pan/tilt camera movements, a motor neuron controlling the magnification of the zoom and an actuator neuron which triggers the action of selecting the targeted stimulus. The basic architecture we tested is a feedforward network, though in some experiments we added recurrence at the hidden layer. Also, in some conditions we added the motor efferences to the input layer.

The Adaptive Algorithm. To establish the connection weights of the neural network we used a genetic algorithm that permits to evolve a population of neuroagents. In a given population of 100 agents, all robots start with random network weights and then they are repeatedly tested on the specified task; during this test set, every performance is recorded and evaluated in terms of fitness. The fitness function is a value that quantitatively estimates the behavior of an evolving robot in a given task (i.e.: exploration, targets discrimination, predator recognizing). The values of the fitness are compared and the best 20 robots will be selected to recreate a whole new population. Each of the selected robots will generate an offspring of 5 agents with a mutation rate of the connection weights and biases of 2 %. These new generation robots will constitute a new population and will be tested on the same tasks; their fitness will be compared and they will undergo the same cycle of selection. This Darwinian process is iterated for 1000 generations.

2.2   The Task and the Experimental Setting

To induce and evaluate the conditioned fear, we built up an experimental setting in which the robot had to perform a well-known neuropsychological test, the cancellation task. The cancellation task is a widely used test and generally consists of asking the subjects to mark with a sign (e.g. using a pen or a finger) random positioned stimuli on a sheet; stimuli can be of different in shape (e.g. dots, circles, lines), colour or quantity and the indications may vary according to the specific aim of the test. Subjects can be asked to mark all the stimuli, only those in a specific position or just a part of them. This task, performed by human beings, is generally used for the diagnosis and discrimination of several kinds of spatial exploration and attentional deficits. In our experiments, the visual scene contains randomly scattered dots of two different colours: 50 % are light red (high luminance stimuli) and 50 % dark red (low luminance stimuli).

Fig. 1. Schematic representation of the experimental setting (Color figure online)

The number of stimuli presented changed for each trial and could vary among 2, 4, 8 or 16. When a stimulus is touched by the robot, it is cancelled from the visual scene. The neuroagents' performance is based on the number of cancelled stimuli; in half of the trials, however, light red stimuli become aversive and then caused the robot's end of life for the specific trial. The total number of possible trials combinations is therefore 8 (4 for each condition). The change of condition is signaled by an external input (in our case, the lighting of a lamp);

an example of the condition can be found in Fig. 1. The artificial agents were evolved to distinguish between the two conditions and to select/cancel all stimuli in absence of the input (no-stress condition), and only dark red targets in the other case (stressful condition).

3   Experiments

3.1   Determining the Fitness Function

We carried out three main preliminary experiments in order to set up a stable environment for our platform. In the first experiment, our aim was to determine which of the used fitness functions allowed the neuroagents to reach the best performance to the task. The architecture tested in this experiment was exclusively feedforward. We compared the performance to the task of two groups of simulated robots:

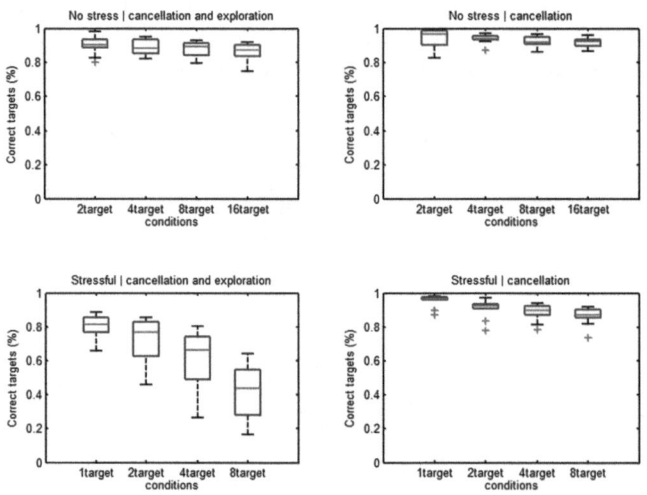

Fig. 2. Boxplots comparing the performances of robots for each fitness function

One equipped with a fitness function which incremented with the robot's exploration ability along with his performance (number of correct targets cancelled), and the second equipped with a function which incremented only for the performance. We evolved 10 different populations of robots for each group. In this training set, every population was evolved for 1000 generations, and every generation was tested on 20 trials, of which half in the stressful condition and the other half in the no-stress condition. For each trial, the robot could move its artificial eye for 1000 steps, at the end of which, the trial was considered finished; if a dangerous stimulus was cancelled during the stressful condition, the trial was terminated immediately. The 10 most evolved robots (i.e. with the highest fitness value) for each group, for a total of 20 individuals, was then subjected to a test set of 1000 trials, and their performances were compared. The results showed that the highest number of correct stimuli were cancelled by the robots of the second group: the robots who were evolved exclusively for their ability to cancel, scored higher than the others, as shown in Fig. 2. In particular, in the last trial of the stressful condition the significance was $p = 0,000183$ scored with the Mann-Whitney U test.

3.2   Balancing the Activation of the Actuator

While analyzing the activation data of the networks' neurons, we noticed that the actuator unit (which triggered the movement of selecting and therefore cancelling stimuli) had an unexpected longer activity in terms of steps; in other words, the number of times the unit was active did not match with the number of the stimuli cancelled by the robots. This meant that the neuroagents, especially in the no-stress condition, tended to cancel target items as well as empty areas on the visual scene; this technique, in fact, allowed them to reach a higher performance. In order to punish the emergence of this sub-optimal behavior, we decided to dramatically reduce the amount of the robots' available steps in a trial in case they cancelled an empty area. We built up two experimental conditions: in the first one, if the actuator unit was active on an empty area, the robot's lifecycle was reduced of 250 steps, in the second one, the lifecycle was reduced of 1000 (the trial was terminated). The neural architecture used was exclusively feedforward and the fitness function incremented along with the only performance, according to the results of Experiment 1. For both conditions, we evolved for the training set 10 populations of robots for 1000 generations on 20 trials each. After the evolution, the best 10 individuals for each condition was tested on a set of 1000 trials, and results are showed in figure. There was a significant evidence that the robots within the first group (which were punished with a reduction of only 250 steps) had a better performance than the second group in the no-stress condition (respectively $p = 0,000999$, $p = 0,000502$, $p = 0,000245$, $p = 0,000187$ for the 2, 4, 8, 16 target no-stress trials). Results are showed in Fig. 3.

3.3   Comparing Architectures

Once all the other variables were settled, it was needed to establish which of the neural architectures best fitted with our emotional task. In this last experiment, we compared the performances of four different networks: (I) a feedforward architecture; (II) an architecture with motor efferences; (III) a recurrent network; (IV) a recurrent network with motor efferences. The fitness was evaluated on the number of correct targets cancelled and if the actuator selected an empty area the trial steps were reduced of 250, in concordance with the results of the first and second experiment.

According to Panksepp's model, as described above, primary and secondary processes arise from the most ancient parts of the brain and thus without involving the cortex; these affective states stems from the brainstem and proceed through the limbic system, and do not need cognition or higher processing to be generated. Both recurrent architectures and networks provided with motor efferences have a high level of information encoding and a memory of the movements in the previous steps. These additional connections, when active, require

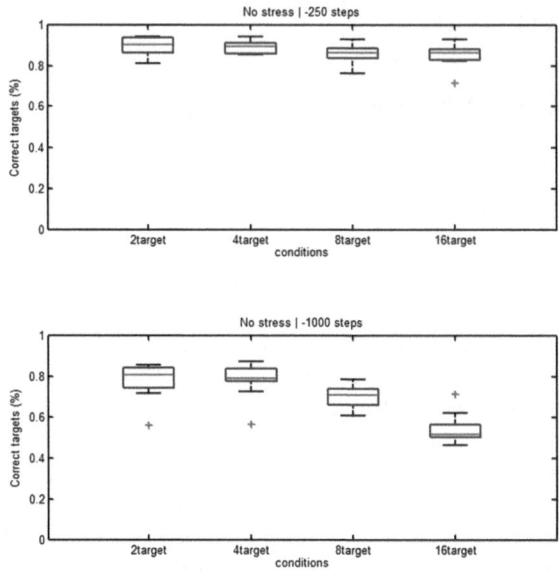

Fig. 3. Boxplots comparing the performances of robots in the no-stress condition punished with a reduction of 250 or 1000 steps in case of cancellation of an empty area.

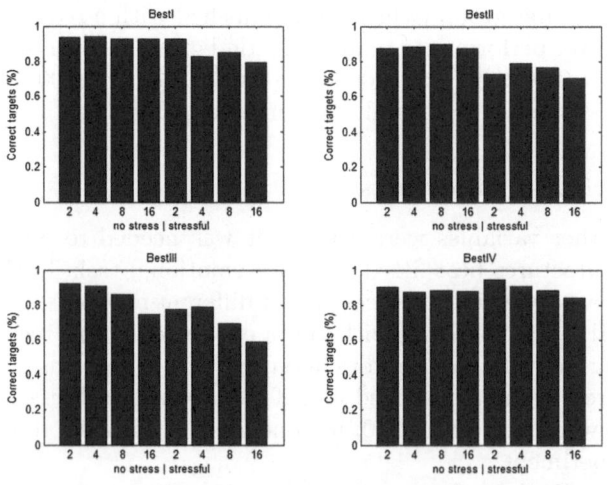

Fig. 4. Correct targets detected by the best robot for each architecture

more time for the computation of a single step and the detection of dangerous targets could be delayed; for this reason, we expect the simplest circuit, the feed-forward network, to show better reaction and to have a better performance than all the others. We trained 10 populations of robots for each architecture for 1000 generations, each lasting 20 trials. The 10 best individuals for each population a total of 40 individuals - were tested on 1000 trials and their performances in terms of fitness were compared. The Mann Whitney U test showed no significant difference among them (p < 0.05).

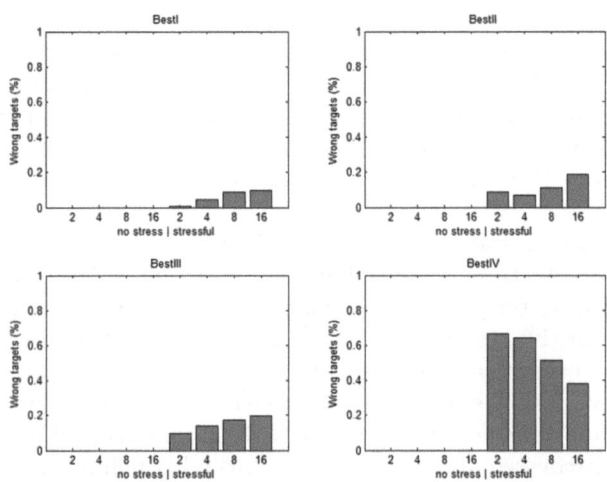

Fig. 5. Wrong targets detected by the best robot for each architecture

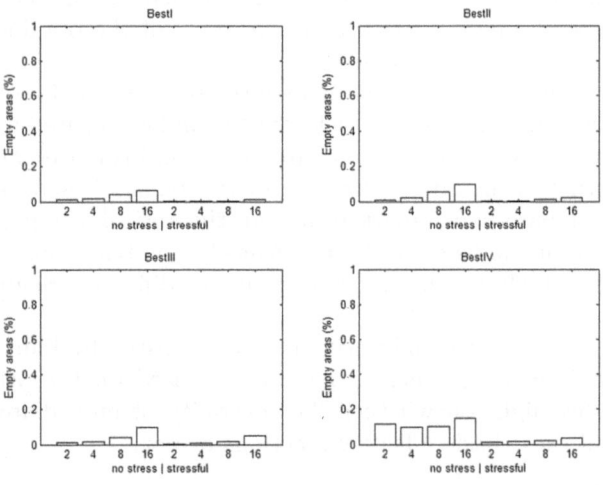

Fig. 6. Empty areas detected by the best robot for each architecture

We conducted further analyses extracting the single best among the best 10 individuals of each group, which we will call BestI, BestII, BestIII and BestIV for the respective architectures. We left them free to move for 1000 trials without any steps reduction and then compared their performance on correct targets, wrong targets and empty areas cancelled.

Graphics show that the better performance was reached by BestI and the difference is highly significant (Figs. 4, 5 and 6).

4   Summary and Direction for Future Researches

We described a method to investigate the adaptive function of ancestral fear and trained artificial simulated humanoid agents to discriminate stressful and no-stress conditions in order to organize an efficient behavior and maximize their survivability. Our platform allows a rapid porting of the simulations on the physical robot iCub and therefore classifies as a hybrid software-hardware system.

Our preliminary results showed the efficiency of the simplest among the 4 tested neural architectures, confirming the importance of a rapid encoding of information for a better performance and adaptability to a given environment. Emotions and affective states provide natural agents with this literally immediate processing and reaction as soon as - like in our case - a danger is perceived, and therefore we feel like giving a contribution in favor of the theses on the phylogenetic importance of emotions. Regarding our artificial neuroagents, still much must be discovered about the true reasons lying behind their performance, and only a qualitative analysis of their behavior and movements, as well as their use of the zoom function, will reveal the difference among the results.

Other open questions regard the possibility of cutting the connections of the network with the emotion unit to watch the behavioral change of the best individuals tested. Would this result in a reversal of the obtained results? Or would the proportions be maintained?

The aim of future researches will include the replication of a standard experiment setting with various kinds of tasks which can be executed by both natural and artificial agents in order to compare their performances and to validate our network structure based on the described models of emotions. Regarding the experimental paradigm, apart from our theoretical perspective, our tool constitutes a robotic model of the human cancellation task and its future implementation could include the reproduction of different neuropsychological phenomena.

Finally, FEAR is not the only primary process listed by Panksepp, and further simulations including different paradigms based on the other 6 primary emotions will shed light on whether the centrality of immediate processing is confirmed for all these basic affective reactions or not.

References

1. Bechara, A., Tranel, D., Damasio, H., Adolphs, R., Rockland, C., Damasio, A.: Double dissociation of conditioning and declarative knowledge relative to the amygdala and hippocampus in humans. Science **269**, 1115–1118 (1995)
2. Canamero, L.: Emotions understanding from the perspective of autonomous robots research. Neural Netw. **18**, 445–455 (2005)
3. Di Ferdinando, A., Parisi, D., Bartolomeo, P.: Modeling orienting behavior and its disorders with ecological neural networks. J. Cogn. Neurosci. **19**, 1033–1049 (2007)
4. Gigliotta, O., Bartolomeo, P., Orazio, O.: Introducing sensory-motor apparatus in neuropsychological modelization. In: Artificial Intelligence and Cognition, p. 80 (2013)
5. James, W.: On some hegelisms. Mind **7**(26), 186–208 (1882)
6. Kolb, B., Tees, R.: The Cerebral Cortex of the Rat. MIT press, Cambridge (1990)
7. LeDoux, J.: Emotional networks and motor control: a fearful view. Prog. Brain Res. **107**, 437–446 (1996)
8. LeDoux, J.: The emotional brain, fear and the amygdala. Cell. Mol. Neurobiol. **23**, 727–738 (2003)
9. Metta, G., Sandini, G., Vemon, D., Natale, L., Nori, F.: The icub humanoid robot: an open platform for research in embodied cognition. In: PerMIS '08 Proceedings of the 8th Workshop on Performance Metrics for Intelligent Systems, pp. 50–56 (2008)
10. Nolfi, S., Floreano, D.: Evolutionary Robotics: The Biology, Intelligence, and Technology of Self-Organizing Machines. MIT Press, Cambridge (2001)
11. Nolfi, S., Gigliotta, O.: Evorobot: a tool for running experiments on the evolution of communication. In: Nolfi, S., Mirolli, M. (eds.) Evolution of Communication and Language in Embodied Agents, pp. 297–301. Springer, Berlin (2010)
12. Panksepp, J.: Affective Neuroscience: The Foundation of Human and Animal Emotions. Oxford University Press, New York (1998)
13. Panksepp, J.: Simulating the primal affective mentalities of the mammalian brain: a fugue on the emotional feelings of mental life and implications for AI-robotics. In: Dietrich, D., Fodor, G., Zucker, G., Bruckner, D. (eds.) Simulating the Mind: A Technical Neuropsychoanalytic Approach, pp. 149–177. Springer, New York (2008)
14. Panksepp, J.: Affective consciousness in animals: perspectives on dimensional and primary process emotion approaches. Proc. R. Soc. **277**(1696), 2905–2907 (2010)
15. Panksepp, J.: Cross-species neuroaffective parsing of primal emotional desires and aversions in mammals. Emotion Rev. **5**(3), 235–240 (2013)
16. Panksepp, J., Fuchs, T., Iacobucci, P.: The basic neuroscience of emotional experiences in mammals: the case of subcortical fear circuitry and implications for clinical anxiety. Appl. Anim. Behav. Sci. **129**(1), 1–17 (2011)
17. Panksepp, J., Biven, L.: The Archaeology of Mind: Neuroevolutionary Origins of Human Emotions. W. W. Norton & Company, New York (2012)
18. Parisi, D., Petrosino, G.: Robots that have emotions. Adapt. Behav. **18**(6), 453–469 (2010)
19. Schmitz, A., Pattacini, U., Nori, F., Natale, L., Metta, G., Sandini, G.: Design, realization and sensorization of the dexterous icub hand. In: International Conference on Humanoid Robots, pp. 186–191 (2010)

Approaches to Molecular Communication Between Synthetic Compartments Based on Encapsulated Chemical Oscillators

Pasquale Stano[1], Florian Wodlei[2], Paolo Carrara[1], Sandra Ristori[3],
Nadia Marchettini[4], and Federico Rossi[5]([⊠])

[1] Science Department, RomaTre University, V.le Marconi 446, 00146 Rome, Italy
[2] Living Systems Research, Roseggerstr. 27/2, 9020 Klagenfurt, Austria
[3] Dipartimento di Scienze della Terra & CSGI, Università degli Studi di Firenze,
Via della Lastruccia 3, 50019 Sesto Fiorentino, FI, Italy
[4] Department of Earth, Environmental and Physical Sciences - DEEP Sciences,
University of Siena, Pian dei Mantellini 44, 53100 Siena, Italy
[5] Department of Chemistry and Biology, University of Salerno,
Via Giovanni Paolo II 132, 84084 Fisciano, SA, Italy
frossi@unisa.it

Abstract. The use of confined micro-oscillators as paradigmatic model for studying communication and information exchange among network of synthetic cells is rapidly growing in the last years. In this paper we report the first steps of an ongoing investigation on the encapsulation of the Belousov-Zhabotinsky oscillating reaction inside phospholipid vesicles (liposomes). The preparation of liposomes encapsulating water soluble molecules can be efficiently carried out in two steps: 1. confining the solute inside water-in-oil droplet, 2. transformation of droplets into liposomes. Here we have started the investigation of chemical oscillation emerging behavior in these two types of compartments. We firstly show interesting dynamical behavior within emulsion droplets. Next, we assess the influence of additives (sugars in particular, which are necessary for the liposomes production) on the oscillation pattern. Future studies will be devoted to the encapsulation of Belousov-Zhabotinsky reaction within liposomes. The potentiality of our systems for modeling intercellular communication pathways and its applications in the Bio-Chem-ITs context are shortly discussed.

Keywords: Chemical communication · Belousov-Zhabotinsky reaction · Lipid vesicles · Synthetic cells

1 Introduction

The construction of synthetic cells, intended as simplified models of biological cells, is one of the cutting-edge projects in synthetic biology. Being focused on the constructive paradigm, cell models of minimal complexity helps understanding the physico-chemical basis of cellular life via a synthetic (putting together)

© Springer International Publishing Switzerland 2014
C. Pizzuti and G. Spezzano (Eds.): WIVACE 2014, CCIS 445, pp. 58–74, 2014.
DOI: 10.1007/978-3-319-12745-3_6

process, rather than following the classical analytical one (taking apart). In the synthetic approach, our understanding of a process/system is compared with (and possibly generated by) our ability to reconstruct it (for example, a metabolic cycle, a regulative circuit). Synthetic cells are constructed by encapsulating molecular components inside lipid vesicles (Fig. 1a), and most of the current research is somehow linked to the origins of life, namely to investigate how primitive cells originate from inanimate matter, and what are their properties. Synthetic cells, however, are useful not only in origin of life research but also for exploring biological complexity in a bottom-up fashion, and for developing novel biotechnological tools.

Among different approaches, the so-called *semi-synthetic* [18,24] is one of the most promising, because it is based on modern macromolecules like enzymes, nucleic acids, ribosomes, etc. that can perform sophisticated functions owing to the fact that their performances have been shaped by billion years of evolution. It has been recently proposed that semi-synthetic cells of minimal complexity can be designed and constructed in order to be able to communicate between each other and with natural cells (Fig. 1b), by using chemical signals [20,25]. This is in principle possible because the current technology should allow the production of synthetic cells endowed with receptors, transcriptional regulators, signal synthases, etc. This research will lead to the construction of "smart" synthetic cells capable of interfacing with biological cells, for example to be used in nanomedicine [16]. It is useful to remark that this scenario has been often described as bio-chemical Information Technologies Bio-Chem-ITs [2], because it is ultimately based on chemical information processing capability of individual molecules or multi-molecular networks, devices and systems.

In the Bio-Chem-ITs context, an intriguing open question is whether basic molecular communication mechanisms, which could resemble those occurring in biological cells, can be achieved by purely chemical systems, without the need of sophisticated biological macromolecules. These chemical communication mechanisms should rely uniquely on the physics and chemistry of simple chemical networks, and their properties (which could also be quite complex) should emerge from a combination of reactivity, compartmentalization and diffusion. To face this problem, we started a thorough investigation by modeling intercellular communication by means of a chemical system encapsulated in liposomes through microfluidic techniques [21,36]. This approach complements the semi-synthetic one and both pave the way to a wide, interdisciplinary and innovative approach for exploiting the computation potentiality of synthetic cells.

In this paper we contribute to the ongoing research [21,29,30,36] on chemical synthetic cell models designed for studying the exchange of chemical signals across phospholipid membranes and the resulting communication patterns. In particular, we encapsulate a chemical oscillator, namely the Belousov-Zhabotinsky (BZ) reaction, into liposomes in order to obtain a network of confined signals emitters/receivers, able to selectively exchange information. Respect to other currently explored encapsulation strategies, such as those based on microfluidics, here we propose a simple and straightforward method to confine the BZ reaction in

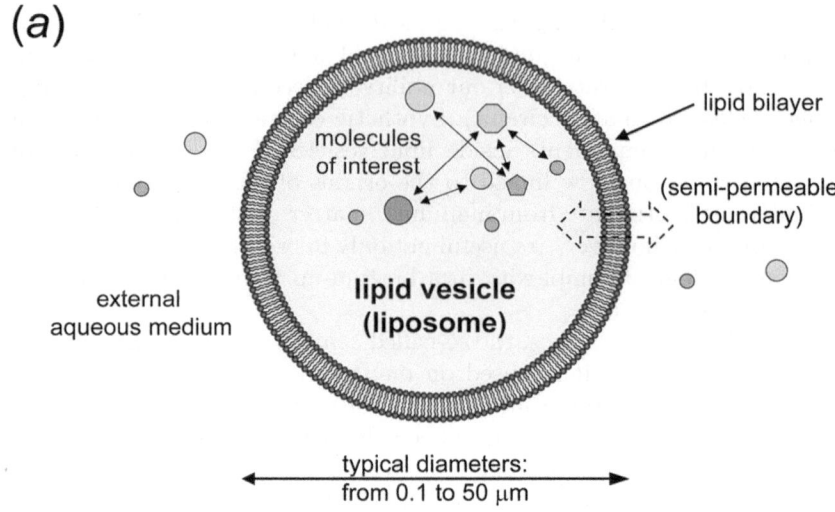

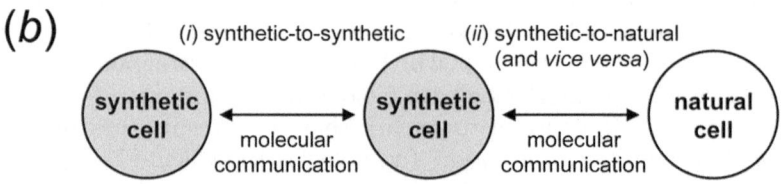

Fig. 1. Synthetic cells and molecular communication. (a) Synthetic cells can be generated by encapsulating a set of chemicals inside a compartment. Most used compartments are lipid vesicles (liposomes) which are characterized by a semi-permeable membrane boundary consisting in a self-assembled lipid bilayer. Although liposomes are prepared in aqueous media, and therefore share the same architecture of biological cells, their size can vary of about two order of magnitude (i.e., from 0.1 to 50–100 µm). (b) Molecular communication is an essential feature of biological cells. Can synthetic cell also communicate via chemical exchanges? This is the topic of a recently started program (that we called "synthetic molecular communication") aimed at constructing semi-synthetic minimal cells, based on proteins and nucleic acids, that can communicate with each other and with natural (biological) cells.

micrometer-size compartments. We will show step-by-step the encapsulation strategy, reporting the interesting dynamical behavior of emulsion droplets obtained as intermediary products of synthesis process.

The following sections will briefly introduce the basic features of confinement and of chemical oscillators. Later in the text we will describe preliminary results and comment on the next developments.

1.1   Soft Matter Confinement

Life is based on biological cells, which are microcompartments where biological molecules are confined and concentrated, whereas the whole cell acts essentially as a reactor enclosed in a semi-permeable boundary. Fascinated by biological architecture, the artificial compartmentalization of chemical and biological species into micro- and nano-structured systems with defined geometry has attracted increasing interest in the latest decades. Confined systems have been used as cellular models, or for a variety of purposes, including the fabrication of advanced materials, for sensing applications or drug-delivery purposes. In most cases, confinement is obtained thanks to the spontaneous formation of compartments based on boundaries composed by lipid molecules (or, more generally, amphiphilic molecules). Amphiphiles that self-assembly in oriented mono- and bi-layers (lamellae) can form micelles, vesicles, cubic phases, and can stabilize biphasic emulsions. Two of the most popular compartments that have been used as cellular models are lipid-stabilized water-in-oil (w/o) droplets and lipid vesicles (liposomes), like those reported in Fig. 2. In particular, the use of w/o droplets and vesicles with dimension in the 1–100 µm allows their direct observation by optical microscopy and since this scale range is also characteristic of cell size, the use of these compartments is very valuable for cytomimetic studies. The possibility to study each compartment individually, via microscopy, is also attractive because it might reveal intriguing between-compartment heterogeneity that would otherwise be impossible to observe by conventional (averaging) techniques. Moreover, by direct observation it will become possible to study how different compartments interact with each other, revealing potential emergent behavior (e.g., synchronization, collective patterns, long-range order, . . .). Until now, w/o macroemulsion droplets and *giant* vesicles (GVs) have been indeed extensively used as cellular models [14,17,31,37], in particular as hosts of biochemical reactions also because their large internal volume assures the encapsulation of solutes.

In this paper, we will explore both w/o droplets and GVs as artificial cell-size compartments capable of hosting the BZ reaction. This choice derives from two reasons. First, we are interested to compare the behavior of chemical oscillations, including inter-compartment communication in these two systems (taking into account the fact that the w/o droplets are suspended in oil, whereas GVs are suspended in an aqueous medium); second, because w/o droplets are actually precursors of GVs according to the recently reported *droplet transfer* method [19]. This innovative method first takes advantage of the facile compartmentalization of water-soluble solutes in w/o droplets, and then convert the solute-filled w/o droplets into GVs.

1.2   Chemical Oscillators as Signals Processors and Generators

Until few decades ago chemical oscillations were thought to be exotic reactions of merely theoretical interest, now they are being studied by several groups with different scientific expertise, from physics to biology [6,40]. Most of the known chemical oscillators, in fact, share complex and peculiar behaviors with many

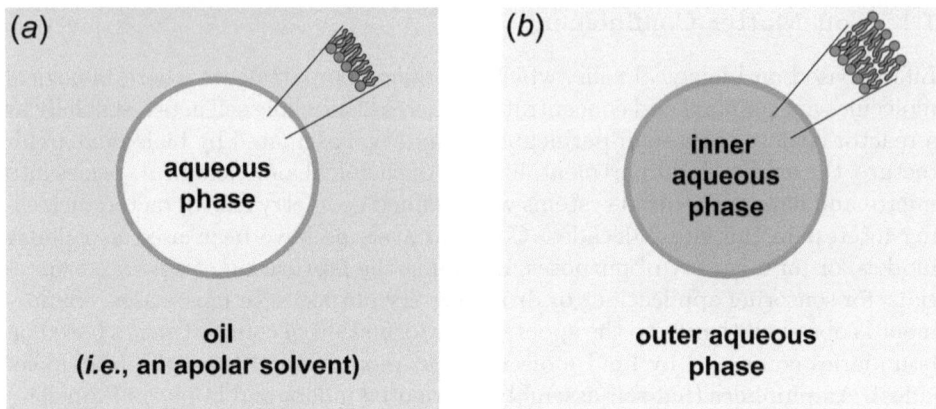

Fig. 2. Comparison between two lipid-based synthetic compartments. (a) Water-in-oil (w/o) emulsion droplets are aqueous droplets suspended in "oil" (*i.e.*, an apolar solvent like hydrocarbons). The interface between water and oil is stabilized by a monolayer of lipids, which are initially dissolved in the oil and self-organize in the presence of water. The w/o droplets are easily obtained by mechanical dispersion of an aqueous phase in the lipid-containing oil. Water-soluble solutes possibly present in the aqueous phase are all found inside the w/o droplets. The typical size of w/o macroemulsion droplets is 1–100 μm. (b) Lipid vesicles, in contrast to w/o droplets, are aqueous compartments suspended in an aqueous phase. The inner and outer phases can be chemically different, and are separated by a lipid bilayer, which is permeable only to small, uncharged solutes. The size of giant lipid vesicles (GVs) is also in the 1–100 μm range. W/o droplets can be transformed in GVs by the droplet transfer method [4,19]. It is remarkable that in this way all solutes initially present in the w/o droplets are found inside GVs.

biological and living systems (excitability, bistability, hysteresis, etc.) and they are regarded as a convenient and simple modeling tool. Among the others, one of the most promising applications of chemical oscillators is their use as "signal transmitter/receiver", where reaction intermediates can be exchanged among different units to probe and study complex communication networks [12,33] or new forms of chemical computing [8,9,11]. The Belousov-Zhabotinsky (BZ) reaction [3,39], in particular, has been used in micro droplets arrays [5,35], lattices [7,34] or bulk systems of catalyst-loaded resin microparticles [32,33]. The dynamical behavior of these systems and/or the pattern selection during the networks evolution is generally determined by the type and magnitude of the coupling strength which is modulated through the exchange of chemical intermediates among individual oscillating units.

The BZ reaction is an oxidation of an organic substrate (malonic acid, MA, in this work) by sodium bromate (Na_2BrO_3) in a sulphuric acid (H_2SO_4) solution and in the presence of a redox catalyst ($Fe(phen)_3^{2+}$, ferroin, in this work). Figure 3 summarizes the basic mechanism responsible for the onset of the oscillations in the concentration of some key intermediates, together with a typical

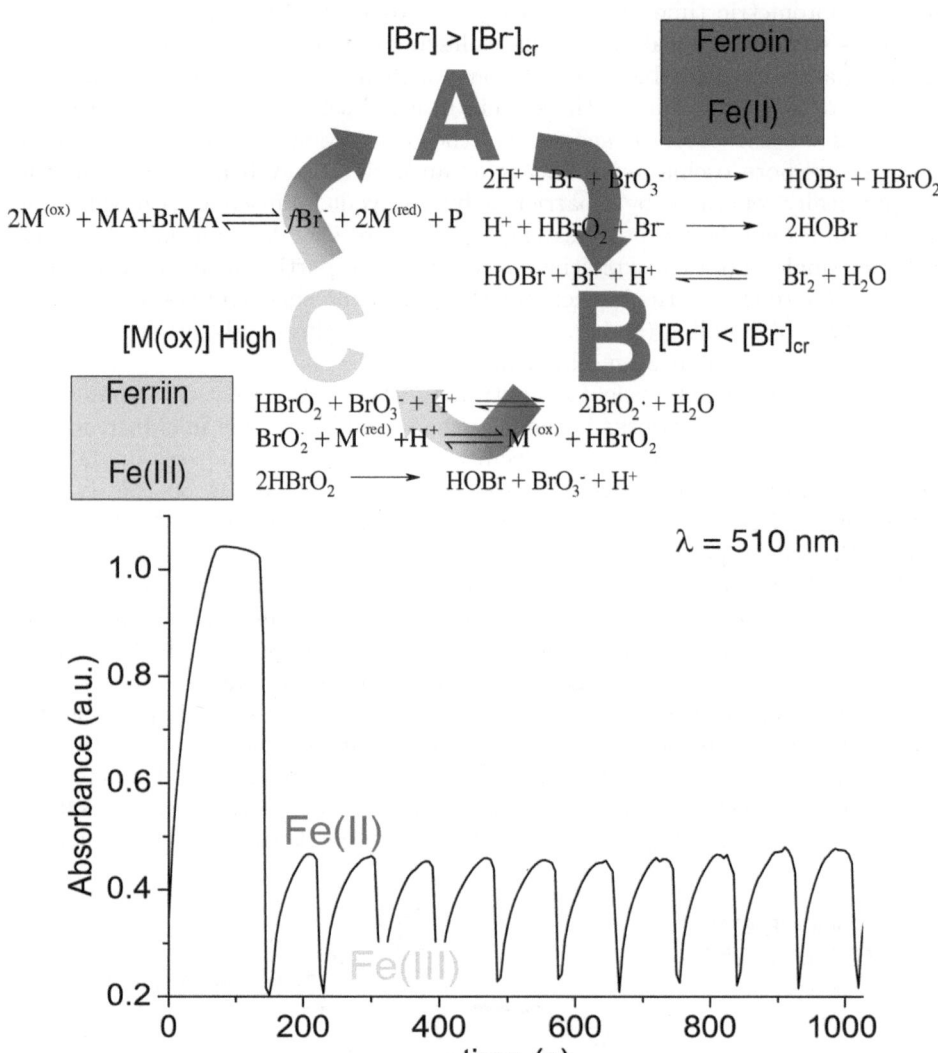

$$[Br] > [Br]_{cr}$$

Ferroin

Fe(II)

A

$$2M^{(ox)} + MA + BrMA \rightleftharpoons fBr^- + 2M^{(red)} + P$$

$$2H^+ + Br^- + BrO_3^- \longrightarrow HOBr + HBrO_2$$

$$H^+ + HBrO_2 + Br^- \longrightarrow 2HOBr$$

$$HOBr + Br^- + H^+ \rightleftharpoons Br_2 + H_2O$$

C

[M(ox)] High

B

$$[Br] < [Br]_{cr}$$

Ferriin

Fe(III)

$$HBrO_2 + BrO_3^- + H^+ \rightleftharpoons 2BrO_2\cdot + H_2O$$

$$BrO_2^- + M^{(red)} + H^+ \rightleftharpoons M^{(ox)} + HBrO_2$$

$$2HBrO_2 \longrightarrow HOBr + BrO_3^- + H^+$$

$$\lambda = 510 \text{ nm}$$

Fe(II)

Fe(III)

Fig. 3. Upper panel schematically describes the oscillating mechanism of the BZ reaction: three key steps are the basic backbone of a much more complex kinetic mechanism, briefly the concentration of the bromide ions (Br⁻, the inhibitor) and the concentration of the oxidized form of the catalyst (ferriin) act as switchers among the three processes, (A) where bromide ions are consumed to yield bromine (Br₂) and bromous acid (HBrO₂, the activator), (B) where the autocatalytic species HBrO₂ oxidizes the reduced form of the catalyst (ferroin) and (C) where the catalyst is reduced by the organic substrate to yield Br⁻ and restart the cycle. The squares in the corners illustrate the colour and redox state changes of the catalyst. Lower panel shows a spectrophotometric time-series at $\lambda = 510$ nm (absorbance maximum of ferroin) in which typical relaxation oscillations show the cycling nature of the reaction mechanism.

spectrophotometric time-series. During each oscillatory cycle, the BZ reaction produces several chemical intermediates having different physico-chemical properties (charge, hydrophobicity, etc.). Some of them act as oscillations inhibitors, like Br^- or Br_2, which bring the system in a reduced state preventing the formation of the autocatalytic species. On the contrary, species like $HBrO_2$ or the catalyst promote oscillations and act as an activator. When single oscillators are physically separated by a barrier or by a solvent immiscible with water, the intermediates can be used as messengers to deliver chemical information among different single units. By exploiting the different properties of the intermediates, it is possible to choose the barriers (or the separating solvent) in order to have a specific type of coupling (global, inhibiting or activating) and the corresponding feedback on the oscillating mechanisms.

Another important issue for controlling the communication dynamics in chemical oscillators networks is the way the signal leaves one unit to reach the others. For example, messenger molecules can target a specific area or they can uniformly spread in the surrounding environment, depending on the dynamical behavior of the chemical oscillator inside a single compartment (see Fig. 4). One of the crucial parameter influencing the global dynamics of the BZ reaction is the medium homogeneity; in particular, in the absence of any macroscopic concentration gradients, the system oscillates uniformly in space (homogeneous oscillations). On the contrary, when spatial gradients are present, the BZ behaves as a reaction-diffusion system generating fronts, pulses or waves. When the BZ system is confined in a vesicle, where no stirring or advection is possible, the critical parameter responsible for the homogeneity of the system is the size of the droplet, i.e. its diameter d. Therefore the spatio-temporal behavior of a confined BZ system drastically depends on the size of the vesicle itself. As sketched in Fig. 4, the type of dynamical behavior inside the droplet dictates the

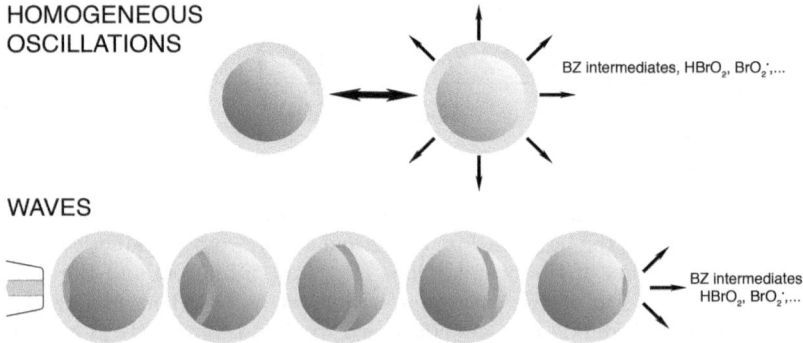

Fig. 4. BZ intermediates (chemical signals) are released from each single vesicle in a fashion that depends on the spatio-temporal dynamics of the confined oscillator. If the system oscillates homogeneously, chemical intermediates do not have any preferential release direction. If a pulse is triggered in a precise point, the signal will follow the direction of the wave propagation to reach a specific target area.

directionality of the signal propagation outside the membranes. Bulk oscillations generate an isotropic distribution of the BZ intermediates in the inner aqueous phase, which, in turn, may cross the amphiphilic barriers in each point of the membrane. Pulses or chemical waves, on the contrary, are directional signals propagating along a straight line, which can be controlled to deliver information to a precise target.

2   Experimental Methods

The BZ reaction was prepared by mixing the following stock solutions : Malonic acid (Sigma), $[MA] = 2\,M$, sulfuric acid (Sigma), $[H_2SO_4] = 5\,M$, sodium bromate (Fluka), $[NaBrO_3] = 5\,M$, ferroin (Fisher), $[Fe] = 25\,mM$. W/o droplets and vesicles were prepared by diluting a stock solution of POPC (Lipoid), $[POPC] = 3\,mM$ in mineral oil (M5904, Sigma). Glucose (Carlo Erba) and Sucrose (Carlo Erba) were used to adjust the density of the solutions for the vesicles preparation (see later in the text for details). All the reagents were of analytical grade and used without further purification.

Figure 5 shows the mechanism of the "droplet transfer" method [4]. As a first step we prepared the inner solution (I-solution) by mixing all (or some) BZ reaction components (Fig. 5a). Next, we disperse the I-solution in 0.5 mM POPC in mineral oil, in order to form the w/o macroemulsion. POPC form a monolayer around w/o droplets, stabilizing the water/oil interface and prevents phase separation and coalescence. The w/o macroemulsion can be used directly to study oscillations within the droplets and droplet/droplet communication. To prepare GVs, we first stratify an interfacial phase, also composed by 0.5 mM POPC in mineral oil, over an aqueous outer solution (O-solution), which should be isotonic with the I-solution. In this way, continuous POPC monolayer self-assemble at the oil/water interface (Fig. 5b). Note that I- and O-solutions are isotonic, but their densities have been adjusted by adding 150 mM sucrose ($\rho \sim 1.24\,g\,cm^{-3}$) to the I-solution and 150 mM glucose ($\rho \sim 1.12\,g\,cm^{-3}$) to the O-solution, such as that $\rho_I > \rho_O$. A freshly prepared w/o macroemulsion is then poured above the two-phase system prepared as described. The emulsion droplets, being denser than oil and denser than O-solution, spontaneously move across the interface, reaching the O-solution. While crossing the interface, a second POPC layer surrounds the droplets, thus forming a bilayer (Fig. 5c). The transfer can proceed by gravity, but it is very slow. We facilitate the droplet transfer by centrifuging the system for 10 min at 5,000 rpm ($\sim$1,700 g). With this procedure, about 30 % of the macroemulsion droplets are generally transformed to vesicles (Fig. 5d).

Investigations on the dynamics of oscillating emulsion droplets were performed in a *pseudo* 2-dimensional reactor. The reactive solution was sandwiched between two 49 mm diameter boro-silicate optical windows, taken apart by a 50 μm thick teflon gasket with an internal diameter of 25 mm. The sample was illuminated from below with a LED light source, while a CCD camera (Imagesource), equipped with a close focus zoom 10 × (Edmund optics) and a band

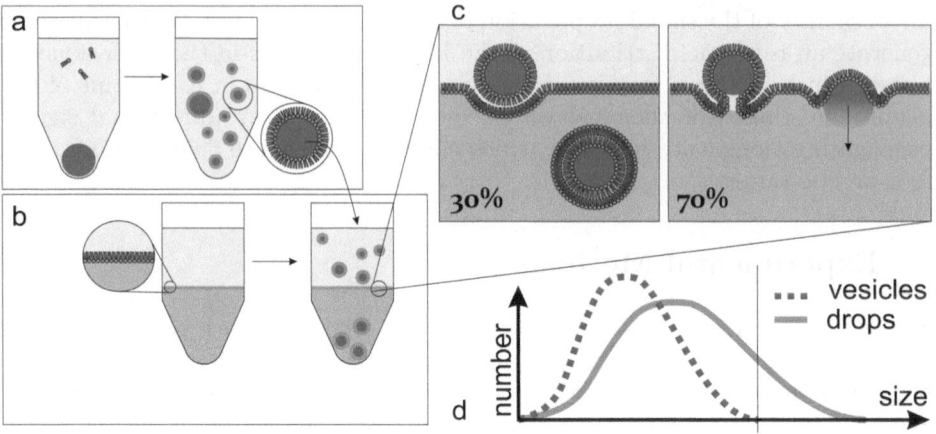

Fig. 5. (a) Preparation of a water-in-oil macroemulsion with mineral oil, surfactants (POPC) and the I-solution. (b) Preparation of an oil over O-solution system with surfactants at the interface. In a second step the macroemulsion droplets are inserted in this system and sink down due to the density difference between the I- and O-solution. (c) As they wander through the interface they get a second layer of surfactants such that they now have a bilayer of phospholipids, i.e. they are vesicles now, if the conditions are good (only in about 30 % of all cases) - otherwise they merge with the interface and the I-solution is released into the O-solution (which is happening in about 70 % of all cases). (d) This leads to a size distribution of the vesicles which does not allow vesicles greater then a critical size, even though the macroemulsion droplets generated in the first step were greater.

pass interference filter (Edmund optics, $\lambda = 510\,\text{nm}$), was placed above the sample and connected to a PC for images acquisition. Space-time (ST) plots for studying the dynamics of oscillating droplets were built by cutting thin slices (1 pixel-thick) from the same region of every frame of the recorded sequences and pasting them sequentially in a new image. Every pixel on the vertical time axis represents the image capture sampling time, and space axis represents the actual space captured by the camera. Time series of the BZ runs were built from the images by extracting the gray level of every pixel, with the highest values (brightest in the images) corresponding to the oxidized form of the catalyst (Fe^{3+}).

3   Results

In this section we describe the results obtained for each step of the "droplet transfer" method. We will first show the successful generation of w/o oscillating droplets. We then illustrate our first preliminary results with vesicles containing only selected BZ reagents. Finally we prove that the addition of the sugars to the oscillating mixture, which is essential to successfully carry out the GVs production, does not influence the qualitative behavior of the system, and, despite few differences, relaxation oscillations in the concentration of the catalyst can still be observed.

3.1 Emulsions

Similarly to previous works by Gorecki and coworkers [1, 29, 30], we prepared a dispersion of a BZ water solution in a oil (mineral oil in this case), stabilized by the presence of phospholipids (POPC in this case). We both explored the case of the homogeneous oscillations inside the droplets and the propagation of chemical pulses among different confined oscillators. Depending on the amount of POPC dissolved in the mineral oil, it is possible to control to a certain extent the number and the size of the aqueous droplets, in particular, at larger amounts of POPC their number increases and their size decreases. Also the intensity of the emulsification process, i.e. stirring and shacking the solution, can influence these two parameters in the same fashion of the lipid concentration.

Figure 6 shows an experiment where oscillating droplets of different sizes (diameter, $d < 100\,\mu\mathrm{m}$) are dispersed in a solution of $[\mathrm{POPC}]_{\mathrm{oil}} = 3\,\mathrm{mM}$. In the panels (a)–(c) three different snapshots illustrate the changes in the color of the catalyst during the oscillatory cycles, from the reduced form (Fe^{2+}, dark color) to the oxidized form (Fe^{3+}, bright color). The behavior of the droplets 1–7 in panel (a), is depicted in the ST plot reported in panel (d) where the bright flashes correspond to the autocatalytic oxidation of the catalyst. From the ST plot it is possible to extract the time series of each single oscillator, like the one referred to the droplet 1 and reported in the panel (e). The analysis of this type of experiments reveals important information about the behavior of the BZ reaction when confined. In particular, the size of the droplets seems to influence both the total lifespan of the oscillations and their period. Smallest droplets ($d < 40\,\mu\mathrm{m}$) after an initial oxidation of the catalyst, do not oscillate along the whole course of the experiments. Droplets that overcome a critical size oscillate in time, but with a period, τ, that depends on d: largest droplets (e.g. 1–4) oscillate faster ($d < 80\,\mu\mathrm{m}$, $\tau \simeq 400\,\mathrm{s}$), smaller droplets (e.g. 5–7) show a longer period ($80 < d < 100\,\mu\mathrm{m}$, $\tau \simeq 500\,\mathrm{s}$). This behavior is probably due to the relative concentration of the reaction products, that depends on the volume of the droplets [28].

When the concentration of the lipid in the mineral oil is smaller and the emulsification process is less intense, it is possible to obtain very large droplets (up to $1000\,\mu\mathrm{m}$) for studying the propagation of waves and chemical pulses. Figure 7(a)–(c) reports the transmission of a pulse from droplet 1 to droplet 4, passing through droplets 2 and 3, when $[\mathrm{POPC}]_{\mathrm{oil}} = 1.5\,\mathrm{mM}$. Note that when two droplets are in close contact with each other, a lipid bilayer separates the two droplets. A chemical pulse is generated in the largest droplet 1 (bright color in panel (a) and transmitted to the droplets 2 and 3, which change their color from dark to bright in panel (b) and finally delivered to droplet 4 (panel c). The transmission of subsequent signals is also evident from the ST plot in panel (d), where horizontal lines represent chemical waves which cross the border of adjacent droplets. The frequency of transmission is halved between droplets 2

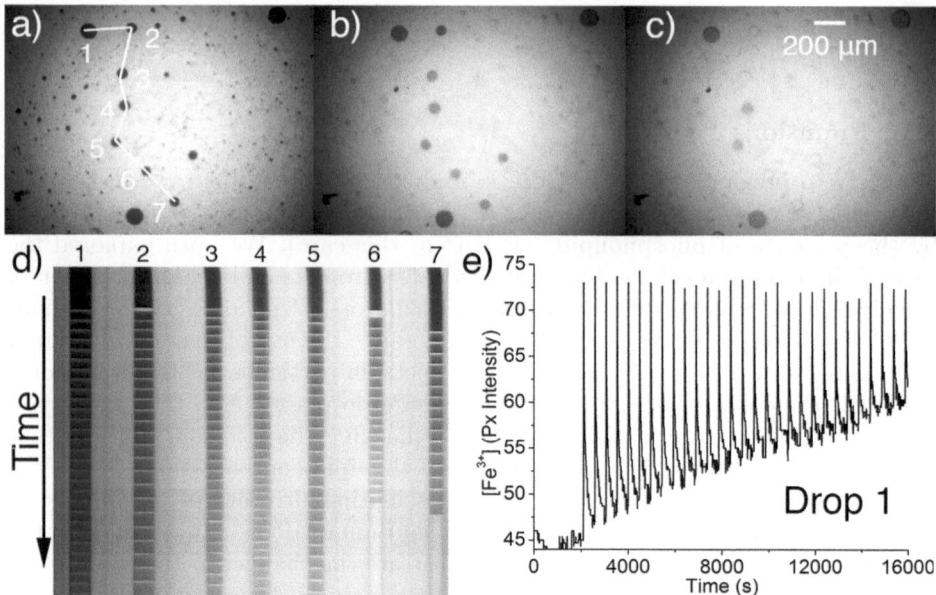

Fig. 6. (a)–(c) Dispersion in mineral oil of BZ aqueous droplets surrounded by lipids. Dark color and bright color indicate the reduced and the oxidized form of the catalyst, respectively. Snapshots are taken about 5000 s apart. (d) ST plot built along the white line in panel (a) showing the oscillations of the droplets 1–7, total time ∼ 16000 s. (e) Time series reporting the oscillatory dynamics of droplet 1 extracted from the ST plot in panel (d). Initial concentrations of the BZ reagents in the water phase are [MA] = 100 mM, [H₂SO₄] = 80 mM, [NaBrO₃] = 300 mM and ferroin = 3 mM. The concentration of the lipid in the oil is [POPC] = 3 mM.

and 3, because of the time needed by droplet 3 to reset the cycle and recover from a refractory to an excitable state. The transmission of chemical impulses indicates that one (or more) reaction intermediates, likely the autocatalytic species $HBrO_2$ as demonstrated by microscopic electrochemical investigations [36], can cross the phospholipid barriers and act as messenger molecule. This behavior was previously demonstrated for the case of activatory coupling in an array of oscillatory droplets generated by means of microfluidic techniques [21].

3.2 Vesicles

As a preliminary step we could successfully produce stable vesicles containing one or more BZ reagents. Figure 8 shows a dispersion of vesicles containing $[BrO_3^-] = 600$ mM and [Fe] = 3 mM after 6 h from their preparation. The outer water solution contains $[BrO_3^-] = 600$ mM to balance the osmotic pressure (the osmotic pressure of 3 mM ferroin is considered negligible in our conditions), and prevent solute leakage via osmotic-induced membrane defects or vesicle damage. At present we are still working to obtain oscillating liposomes containing a

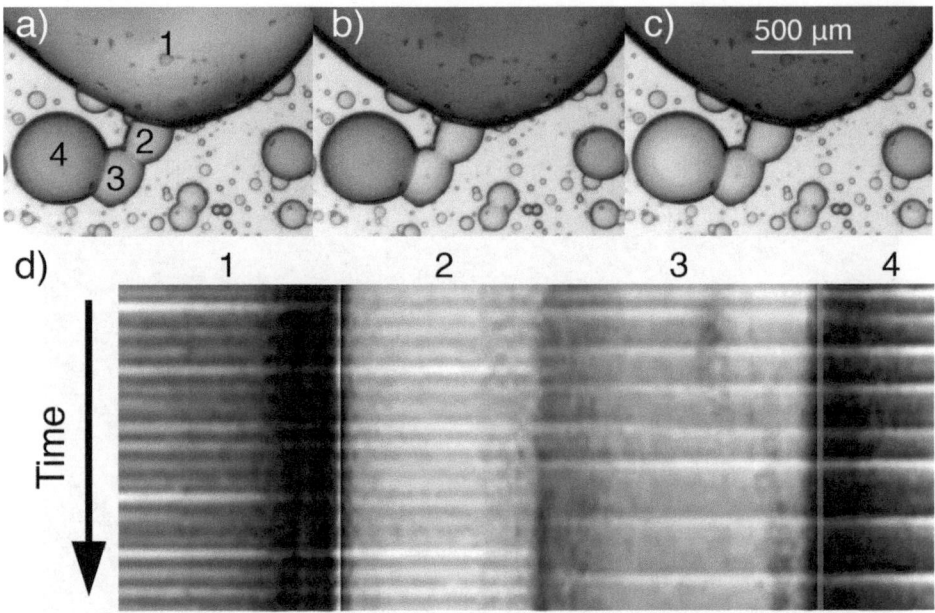

Fig. 7. (a)–(c) Pulse transmission from droplet 1 to droplet 4. (d) ST plot of droplets 1–4, total time $\sim$ 10000 s. Initial concentrations of the BZ reagents in the water phase are $[MA] = 100$ mM, $[H_2SO_4] = 80$ mM, $[NaBrO_3] = 300$ mM and $[Fe] = 3$ mM. The concentration of the lipid in the oil is $[POPC] = 1.5$ mM.

full BZ reaction, through the droplet transfer method. Preliminary experiment revealed that in order to obtain GVs containing all components of BZ reactions a fine tuning of the physical and chemical experimental parameters is required. However, since we already demonstrated that stable vesicles can be generated by means of microfluidic techniques [21,36] reaching our goal seems an achievable task. The most critical parameter which seems to affect the vesicle stability is the low pH typical of the BZ reaction. However, other processes, like the carbon dioxide production during the oxidation of the organic substrate, may contribute to the membranes instability. Possibly, the use of specific lipid or lipid mixtures could improve the efficiency of GVs formation. The exploration of the best conditions to obtain BZ-reaction-containing GVs by the droplet transfer method is currently under exploration in our laboratories.

3.3   Influence of the Added Sugars on the BZ Reaction

As explained in the experimental section, one of the key feature of the droplet transfer method is the need of creating a sugar-based density gradient (I-solution vs. O-solution) that facilitate the droplet-to-GVs transformation. We then performed a series of experiments in order to check the influence of sugars, i.e. oxidizable substrates, on the oscillating mechanism of the BZ reaction. Sevcik

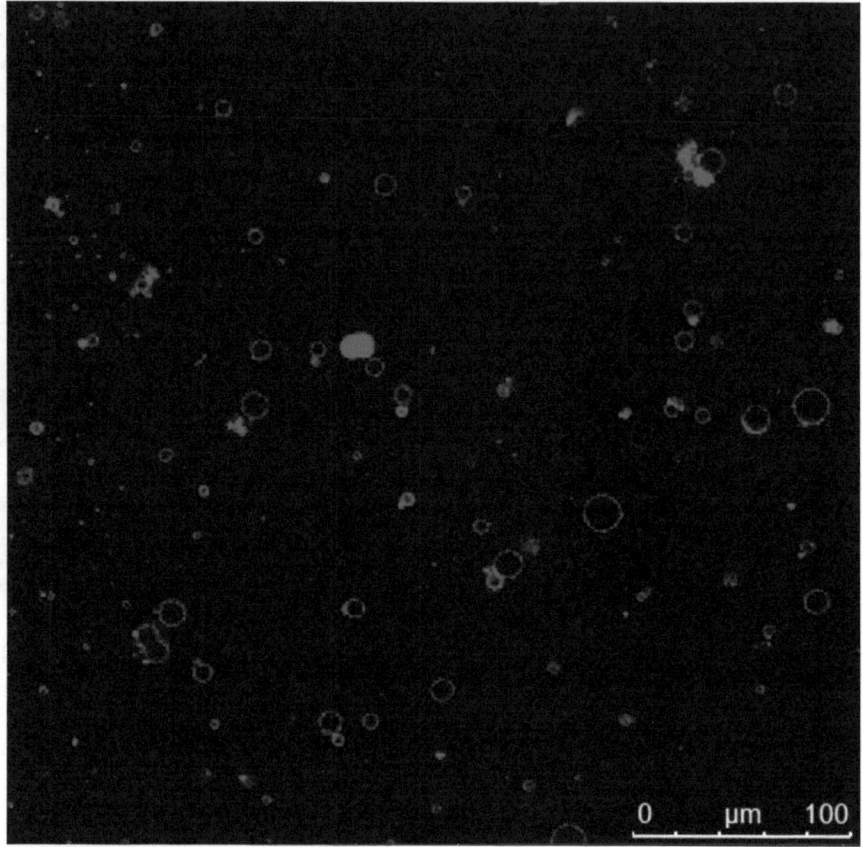

Fig. 8. Microscope picture of the vesicles. Red circlets indicate stable vesicles containing $[BrO_3^-] = 600\,mM$ and $[Fe] = 3\,mM$ after 6 h from their preparation. The vesicles size ranges from about $5\,\mu m$ to $20\,\mu m$.

et al. [22,23] investigated the temporal behavior of the BZ reaction when cerium sulfate was used as the catalyst and glucose as the organic substrate in the place of malonic acid. However, with ferroin as the catalyst, the system has never been characterized. In order to decouple the effect of confinement from the effect of the substrate, we explored several BZ/sugar systems in water solution, with increasing concentrations of glucose (GLU) and sucrose in the presence of MA. The two additives were found to have similar effects on the global dynamics of the BZ reaction; as an example Fig. 9 shows the comparison between a BZ reaction in the absence (panel a) and in the presence of $[GLU] = 5\,mM$ (panel b). The presence of glucose causes the change of the main oscillating parameters, namely a shorter period (τ decreases from ~20 s to ~8 s), a lower amplitude and the appearance of a pre-oscillatory induction period (I.P.), which length depends linearly on [GLU] (see Fig. 9c). These differences are mainly due to the role of

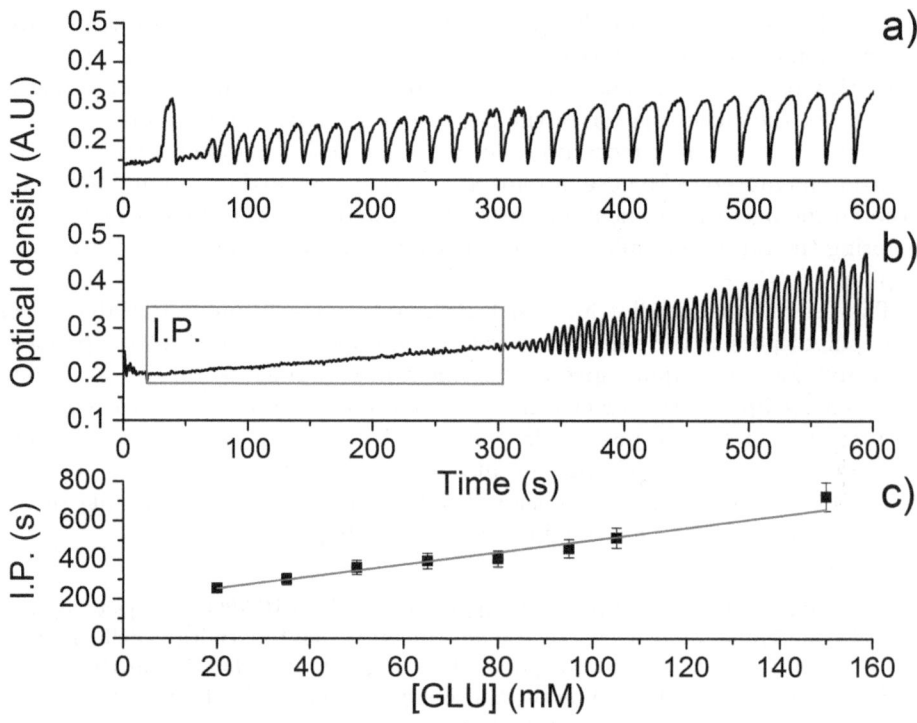

Fig. 9. Absorbance of ferroin at 492 nm with (a) [GLU] = 0 M and (b) [GLU] = 35 mM. (c) Length of I.P. as a function of increasing [GLU]. Initial concentrations of the BZ reagents are [MA] = 37 mM, [H$_2$SO$_4$] = 396 mM, [NaBrO$_3$] = 176 mM and [Fe] = 0.55 mM. In all measurements the system was stirred at a constant rate.

glucose as scavenger for bromine molecules, which results in a lower production of bromomalonic acid as the source for bromide ions; the latter being an important species for the BZ mechanism. However, despite quantitative differences, the overall oscillating mechanism is preserved even in the presence of relatively high concentrations of sugars.

4   Conclusions and Perspectives

Since the pioneering works by Winfree and Kuramoto [15,38] networks of communicating chemical oscillators have been regarded as a powerful tool for understanding the innermost dynamics of biological and biophysical networks. Most of them, from cardiac pacemakers to the neuronal activity, are in fact pulse coupled. Previous works showed that networks of BZ reactions compartmentalized in a water-in-oil emulsion, self-organize in complex dynamical behaviors when coupled through inhibitory chemical species [34,35]. Chemical oscillators also raised a certain interest in the Bio-Chem-ITs context [8], where the BZ reaction has been successfully employed in maze solving problems [26] and in the construction of chemical diodes [10], and - when in coupled networks - in the construction of memory devices [13] and logical gates [27].

In this paper we presented the first moves toward a simple and versatile strategy for the encapsulation of the BZ reaction in giant phospholipid vesicles, a system which yields a more pronounced biomimetic character respect to the water-in-oil emulsion. The use of phospholipids as the barrier forming molecules allows to engineer the permeability respect to different chemical messengers, thus exploring the full potentiality of inhibitory and activatory (or a mix of the two) coupling.

The use of microfluidics for vesicles generation is certainly advisable when a low polydispersity is desired, however the method introduced in this paper is easier and does not require special equipment or expertise. So far, we succeeded in producing lipid based macroemulsion containing a full set of BZ reagents, the first step of the proposed "droplet transfer" method; moreover we could also produce vesicles containing the catalyst and the oxidizing species. The long lifespan of the oscillations in emulsion droplets and the chemical stability of the vesicles are encouraging for the final success of our project.

Acknowledgments. This work was supported by the PRIN2008 project "Synthcell" (2008FY7RJ4) funded by the Italian Ministero dell'Università e della Ricerca (M.I.U.R.). F.R. work was also supported by the grants "Fondo Giovani Ricercatori 2012" and ORSA120144 funded by the University of Salerno (FARB ex 60 %). Daniela Somma (Univ. RomaTre) helped in early experiments with GVs. We are grateful to Pier Luigi Luisi (Univ. Roma Tre and ETH Zurich) for inspiring discussions on synthetic cells of minimal complexity. P.S. and F.R. acknowledge the support through the COST Action CM1304 (Emergence and Evolution of Complex Chemical Systems).

References

1. Adamatzky, A., Holley, J., Dittrich, P., Gorecki, J., Lacy Costello, B., Zauner, K.P., Bull, L.: On architectures of circuits implemented in simulated BelousovZhabotinsky droplets. Biosystems **109**(1), 72–77 (2012)
2. Amos, M., Dittrich, P., McCaskill, J., Rasmussen, S.: Biological and chemical information technologies. Proc. Comput. Sci. **7**, 5660 (2011)
3. Belousov, B.P.: A periodic reaction and its mechanism. In: Sbornik Referatov po Radiatsonno Meditsine, pp. 145–147. Medgiz, Moscow (1958)
4. Carrara, P., Stano, P., Luisi, P.L.: Giant vesicles colonies: a model for primitive cell communities. ChemBioChem **13**(10), 14971502 (2012)
5. Delgado, J., Li, N., Leda, M., Gonzlez-Ochoa, H.O., Fraden, S., Epstein, I.R.: Coupled oscillations in a 1D emulsion of BelousovZhabotinsky droplets. Soft Matter **7**(7), 3155 (2011)
6. Epstein, I.R., Pojman, J.A.: An Introduction to Nonlinear Chemical Dynamics: Oscillations, Waves, Patterns, and Chaos. Oxford University Press, New York (1998)
7. Fukuda, H., Morimura, H., Kai, S.: Global synchronization in two-dimensional lattices of discrete BelousovZhabotinsky oscillators. Physica D: Nonlinear Phenom. **205**(1–4), 80–86 (2005)

8. Gentili, P.L.: Small steps towards the development of chemical artificial intelligent systems. RSC Adv. **3**(48), 25523–25549 (2013)
9. Gentili, P.L., Horvath, V., Vanag, V.K., Epstein, I.R.: Belousov-zhabotinsky chemical neuron as a binary and fuzzy logic processor. Int. J. Unconv. Comput. **8**(2), 177–192 (2012)
10. Gorecka, J.N., Gorecki, J., Igarashi, Y.: One dimensional chemical signal diode constructed with two nonexcitable barriers. J. Phys. Chem. A **111**(5), 885–889 (2007)
11. Holley, J., Adamatzky, A., Bull, L., Lacy Costello, B., Jahan, I.: Computational modalities of belousov-zhabotinsky encapsulated vesicles. Nano Commun. Netw. **2**(1), 50–61 (2011)
12. Horvath, V., Gentili, P.L., Vanag, V.K., Epstein, I.R.: Pulse-coupled chemical oscillators with time delay. Angew. Chem. Int. Ed. **51**(28), 6878–6881 (2012)
13. Kaminaga, A., Vanag, V.K., Epstein, I.R.: A reaction-diffusion memory device. Angew. Chem. Int. Ed. **45**(19), 3087–3089 (2006)
14. Kato, A., Yanagisawa, M., Sato, Y.T., Fujiwara, K., Yoshikawa, K.: Cell-sized confinement in microspheres accelerates the reaction of gene expression. Sci. Rep. **2**, 283 (2012)
15. Kuramoto, Y.: Chemical Oscillations, Waves, and Turbulence. Courier Dover Publications, New York (2003)
16. LeDuc, P.R., Wong, M.S., Ferreira, P.M., Groff, R.E., Haslinger, K., Koonce, M.P., Lee, W.Y., Love, J.C., McCammon, J.A., Monteiro-Riviere, N.A., Rotello, V.M., Rubloff, G.W., Westervelt, R., Yoda, M.: Towards an in vivo biologically inspired nanofactory. Nat. Nano **2**(1), 37 (2007)
17. Luisi, P.L., Walde, P.: Giant Vesicles: Perspectives in Supramolecular Chemistry. Wiley-Interscience, Chichester (2000)
18. Luisi, P., Ferri, F., Stano, P.: Approaches to semi-synthetic minimal cells: a review. Naturwissenschaften **93**(1), 1–13 (2006)
19. Pautot, S., Frisken, B.J., Weitz, D.A.: Production of unilamellar vesicles using an inverted emulsion. Langmuir **19**(7), 28702879 (2003)
20. Rampioni, G., Damiano, L., Messina, M., D'Angelo, F., Leoni, L., Stano, P.: Chemical communication between synthetic and natural cells: a possible experimental design. Electron. Proc. Theor. Comput. Sci. **130**, 1426 (2013)
21. Rossi, F., Zenati, A., Ristori, S., Noel, J.M., Cabuil, V., Kanoufi, F., Abou-Hassan, A.: Activatory coupling among oscillating droplets produced in microfluidic based devices. Int. J. Unconv. Comput. (2014, in press)
22. Sevcik, P., Adamcikova, L.: The oscillating belousov-zhabotinsky reaction with saccharides in batch systems. React. Kinet. Catal. Lett. **33**(1), 4751 (1987)
23. Sevcik, P., Adamcikova, L.: The oscillating belousov-zhabotinsky type reaction with saccharides. J. Phys. Chem. **89**(24), 51785179 (1985)
24. Stano, P., Carrara, P., Kuruma, Y., Souza, TPd, Luisi, P.L.: Compartmentalized reactions as a case of soft-matter biotechnology: synthesis of proteins and nucleic acids inside lipid vesicles. J. Mater. Chem. **21**(47), 1888718902 (2011)
25. Stano, P., Rampioni, G., Carrara, P., Damiano, L., Leoni, L., Luisi, P.L.: Semi-synthetic minimal cells as a tool for biochemical ICT. Biosystems **109**(1), 2434 (2012)
26. Steinbock, O., Toth, A., Showalter, K.: Navigating complex labyrinths: optimal paths from chemical waves. Science **267**(5199), 868–871 (1995)
27. Steinbock, O., Kettunen, P., Showalter, K.: Chemical wave logic gates. J. Phys. Chem. **100**(49), 18970–18975 (1996)

28. Steinbock, O., Muller, S.C.: Radius-dependent inhibition and activation of chemical oscillations in small droplets. J. Phys. Chem. A **102**(32), 6485–6490 (1998)
29. Szymanski, J., Gorecka, J.N., Igarashi, Y., Gizynski, K., Gorecki, J., Zauner, K.P., de Planque, M.R.R.: Droplets with information processing ability. Int. J. Unconv. Comput. **7**(3), 185–200 (2011)
30. Szymanski, J., Gorecki, J., Hauser, M.J.B.: Chemo-mechanical coupling in reactive droplets. J. Phys. Chem. C **117**(25), 13080–13086 (2013)
31. Tawfik, D.S., Griffiths, A.D.: Man-made cell-like compartments for molecular evolution. Nat. Biotechnol. **16**(7), 652656 (1998)
32. Taylor, A.F., Tinsley, M.R., Wang, F., Huang, Z., Showalter, K.: Dynamical quorum sensing and synchronization in large populations of chemical oscillators. Science **323**(5914), 614–617 (2009)
33. Tinsley, M.R., Nkomo, S., Showalter, K.: Chimera and phase-cluster states in populations of coupled chemical oscillators. Nat. Phys. **8**(9), 662–665 (2012)
34. Toiya, M., Gonzalez-Ochoa, H.O., Vanag, V.K., Fraden, S., Epstein, I.R.: Synchronization of chemical micro-oscillators. J. Phys. Chem. Lett. **1**(8), 1241–1246 (2010)
35. Toiya, M., Vanag, V.K., Epstein, I.R.: Diffusively coupled chemical oscillators in a microfluidic assembly. Angew. Chem. Int. Ed. **47**(40), 7753–7755 (2008)
36. Tomasi, R., Noel, J.M., Zenati, A., Ristori, S., Rossi, F., Cabuil, V., Kanoufi, F., Abou-Hassan, A.: Chemical communication between liposomes encapsulating a chemical oscillatory reaction. Chem. Sci. **5**(5), 1854–1859 (2014)
37. Walde, P., Cosentino, K., Engel, H., Stano, P.: Giant vesicles: preparations and applications. Chembiochem **11**(7), 848865 (2010)
38. Winfree, A.T.: The Geometry of Biological Time. Springer, Heidelberg (2001)
39. Zhabotinsky, A.M.: Periodic liquid phase reactions. Proc. Acad. Sci. USSR **157**, 392–395 (1964)
40. Zhabotinsky, A.M., Rossi, F.: A brief tale on how chemical oscillations became popular an interview with anatol zhabotinsky. Int. J. Ecodyn. **1**(4), 323–326 (2006)

Multi-objective Parameter Tuning for PSO-based Point Cloud Localization

Roberto Ugolotti[✉] and Stefano Cagnoni

Department of Information Engineering, University of Parma,
Parco Area delle Scienze 181/A, 43124 Parma, Italy
rob.ugo@ce.unipr.it

Abstract. It has been largely proven that population-based metaheuristics such as Particle Swarm Optimization (PSO) are severely affected by the choice of their parameters.

In this paper, we use a multi-objective parameter tuning method called EMOPaT (Evolutionary Multi-Objective Parameter Tuning) to optimize PSO when dealing with a real-world optimization task: the localization of an object acquired by a laser scanner in the form of a point cloud.

We want to optimize both the time needed to reach a quality threshold and the final alignment between the point cloud and a reference model of the object. Our system is able to generate "fast" and "precise" versions of PSO and, among all the possible configurations which lie between the fastest and the most precise, the ones that give the best trade-offs between precision and speed.

Keywords: Particle Swarm Optimization · Pattern recognition · Multi-objective optimization · Point clouds

1 Introduction

Evolutionary Computation (EC) and Swarm Intelligence techniques have proven to be very successful in several fields. Among the many possible applications, we have shown that these techniques are able to recognize and localize object in images and video sequences [15].

In a previous work [14], we compared the performance of GPU-based implementations of Particle Swarm Optimization (PSO [5]) and Differential Evolution (DE) [12] against Fast Point Feature Histograms (FPFH [11]) in recognizing and localizing an object acquired with a laser scanner in the form of a point cloud. The two metaheuristics proved their effectiveness, especially when dealing with noisy or incomplete data.

In this paper, we try to further improve the results obtained by PSO by using a multi-objective evolutionary method for automatic parameter tuning. Automatic parameter tuning is a procedure in which techniques such as EC-based metaheuristics are employed to automatically set good parameter values

© Springer International Publishing Switzerland 2014
C. Pizzuti and G. Spezzano (Eds.): WIVACE 2014, CCIS 445, pp. 75–85, 2014.
DOI: 10.1007/978-3-319-12745-3_7

to configure a given algorithm. This procedure is especially useful in EC, because usually evolutionary metaheuristics have many parameters which, if not chosen appropriately, may lead to bad results even on simple problems. Many different techniques have been proposed over the years such as hyper-heuristics [2], racing algorithms [7], model-based methods [1] and Meta-Evolutionary Algorithms [4]. Our multi-objective extension of this paradigm has been firstly presented in [13] and is summarized in Sect. 4.

The remainder of the paper is structured as follows. Section 2 presents the general framework within which this work has been developed; Sect. 3 introduces the problem under consideration; Sect. 4 describes the evolutionary method for parameter tuning; Sects. 5 and 6 describe the experimental setup and results and are followed by some final remarks in Sect. 7.

2   Model Based Object Recognition

The approach used in this work to align two point clouds is an application of the general method described in [15], whose goal is to locate objects in static images or track them in video sequences. The general process is based on these steps:

1. A template of the object to recognize is created off-line, and the available range of deformations to which it can be subject is defined. A template is a mathematical description of the object and of all the possible ways in which it may appear in an image;
2. Every time a new image (or frame, in case of a video) is presented in input, the template is rotated and deformed according to the transformations represented by the individuals of a population-based metaheuristic, such as PSO, driven by a fitness function that appropriately describe the match between the model and the target object under consideration;
3. The process stops when the alignment is good enough or the time allowed for performing the task has expired.

In the case presented here, we want to match the pose of a known object (which is part of a database of available models) with a point cloud acquired by a linear scanner, i.e., to find the transformation of the model which best matches the data in the point cloud. Therefore, our template is a point cloud that can only be subject to a rigid transformation, so the search space can be defined only by six degrees of freedom (translations and rotations around the three axes). Figure 1 shows an example of a PSO particle encoding a roto-translation.

Fig. 1. A PSO particle encoding a possible roto-translation of the reference.

More details regarding the specific application considered in this work are presented in the next section.

3   Point Cloud Alignment

Our method is designed for inclusion in an architecture whose goal is to help users to program robotic tasks, dealing with object handling. To reach this goal, a sub-system for object recognition has been developed (see Fig. 2). It receives input data from a high-resolution planar laser scanner mounted on the wrist of a six degrees of freedom robot arm. The estimated accuracy of the whole measurement chain is about 1.5 cm. Data undergo several preprocessing steps that refine the acquisition and are then passed to the PSO-based recognizer, along with a list of models stored in a database. The output of the recognizer indicates which objects are present in the scene and in which pose. A more detailed description of a preliminary version of this system is presented in [10]. In this paper, we will only consider a single object, since our goal here is just to determine the pose of a known object as a real-world task on which EMOPaT's performance can be demonstrated.

The data acquired from the laser scanner is in the form of a point cloud, i.e., a set of three-dimensional points expressed within a certain coordinate system that represents the external surface of an object. Many solutions have been proposed for Point-cloud registration and alignment. Among others, Li et al [6] propose a function based on a Gaussian Mixture distance map optimized by PSO; in [8], registration of partially overlapped point clouds is achieved by estimating their Extended Gaussian Images; and in [16] the relative distance between a triangular mesh and a point cloud is minimized by means of DE.

3.1   Fitness Function

The goal of the task under consideration is to match the newly acquired point cloud (called target) to a reference one. If the two point clouds refer to the same object, which is the case we are considering in this paper, this is just an alignment problem and the fitness function to be minimized takes into consideration

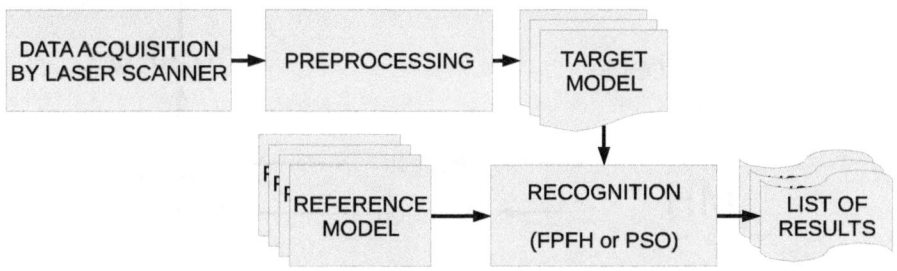

Fig. 2. Representation of the system within which the PSO recognizer is used.

the distance of each point of the target to any of the points that compose the reference. If more models are considered, the same procedure can be used for object classification. Within such a framework, the point cloud can be classified as the object in the database for which this fitness is minimal.

The fitness function used by PSO is relatively straightforward. We compare the target cloud T to be recognized (composed of N_T points), with the reference cloud R, composed of N_R points. This reference is subject to a transformation M encoded by a PSO particle (as in Fig. 1), to obtain a roto-translated version $R' = M(R)$. The fitness of a particle is the average of the minimum distances of each point of T to the closest point of the roto-translated reference R'. More formally:

$$F(T, R') = \frac{1}{N_T} \sum_{p \in T} \min_{q \in R'} \left(dist(p, q) \right)$$

where $dist(x, y)$ is a valid distance metric between two points x and y; in this case we selected the squared Euclidean distance.

Each point cloud is expressed within a local reference frame centered around its centroid. A model can do a full rotation around each axis while the range of translation is limited to 10 cm in each direction, which is good enough to satisfy the requirements of the environment we are considering.

4   EMOPaT

Evolutionary Multi-Objective Parameter Tuning (EMOPaT) was firstly presented in [13]. EMOPaT is a Meta-Evolutionary Algorithm (Meta-EA, see a representation in Fig. 3), i.e. a method in which an Evolutionary Algorithm (in our case NSGA-II [3]) is not used directly to solve a problem, but to tune another metaheuristic (that can be an EA itself), which will then be used to solve the problem.

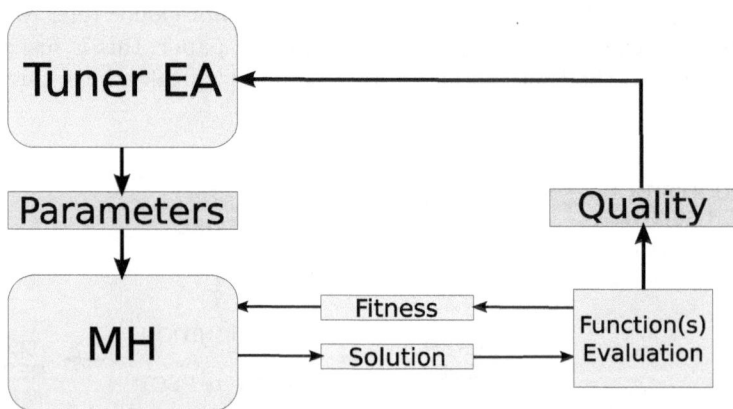

Fig. 3. Scheme of a general Meta-EA, the paradigm on which our tuning method relies.

A Meta-EA (see [4] for a detailed review) works similarly to any other EC technique. It generates a population of possible solutions, each of which is actually an instance of the metaheuristic to optimize. Therefore the tuner EA does not operate in the search space of the problem at hand, but in the search space of the parameters of the metaheuristic considered, which in turn will try to solve the problem.

A Meta-EA tries to find a good set of parameters according to a single objective, usually the fitness function of the EA to be tuned. It is not possible to take into consideration other performance measures at the same time, as, for instance, the time needed to find the optimum. EMOPaT extends classical Meta-EAs by considering more objectives at the same time. This set of objectives (represented by the block "Quality" in Fig. 3) can include any of:

- different functions;
- the same function with different constraints, like time limitations;
- different quality indices used to assess the results of the optimization of a function, such as the fitness value, or the time needed to reach a preset fitness value.

Compared to classical Meta-EA approaches, the result of the process is not only a good set of parameters for solving the problem(s) under consideration, but a population of parameter sets, non-dominated according to the objectives taken into consideration. A parameter set is called non-dominated when no other solution has been found that is better on all objectives. Therefore, each individual in the NSGA-II population represents a good solution, yielding a good trade-off between conflicting goals. The actual result of NSGA-II is then the non-dominated subset of the whole population, which ideally should sample and approximate the Pareto front for the problem at hand, i.e., the whole set of all possible non-dominated solutions.

The analysis of the solutions found by the multi-objective approach to parameter tuning can also provide information on how a parameter affects an EA, how to infer novel solutions from the ones found, and how to generalize them. In [13], by looking at the correlations between parameters and fitness values, we were able to draw same interesting general conclusions about the dependence of the metaheuristics' performance on the value of some parameters, finding, for instance, that the smaller the population, the faster a good solution can be found, while the increase in the number of solutions generally leads to more precise results.

Another interesting ability of EMOPaT is that, unlike classical Meta-EAs, it is able to deal with nominal parameters, such as PSO topology or the kind of crossover used in DE. This is done by reserving in the representation of a NSGA-II individual a real number for each possible choice, and then selecting the parameter setting associated with the highest value.

In [13], EMOPaT has proven its ability in different scenarios, optimizing classical benchmark functions. Its results, besides giving useful hints for understanding parameter semantics, were sets of parameters that compared favorably with other manually- and automatically-tuned sets of parameters. This is the first time we apply EMOPaT to a real-world problem.

5 Experimental Setup

Our goal is to find sets of PSO parameters that are able, according to the needs of the application and the hardware on which it runs, to find a good alignment between point clouds in a limited amount of time. For this reason, the two objectives we let EMOPaT optimize are:

1. Minimum time required to reach a fitness value of 0.1, averaged over 10 repetitions;
2. Best fitness reached after a time limit of 6 s, averaged over 10 repetitions.

Tests were run on a machine equipped with a 64-bit Intel Core i7 CPU running at 3.40 GHz using CUDA v. 5.0 on an nVidia GeForce GTX680 graphics card with 1536 cores working at 1.20 GHz and compute capability 3.0. PSO was implemented on CUDA using the open source library LibCudaOptimize [9] and running on GPU, while NSGA-II was implemented in C++, since the advantages of implementing it on GPU would not have been significant.

NSGA-II was run with these parameters: 60 individuals, 30 generations, mutation rate $= 0.125$, crossover rate $= 0.9$. The ranges within which PSO parameters were allowed to vary are shown in Table 1. Referring to Fig. 3, NSGA-II is the Tuner EA, PSO is the implemented metaheuristic, and instead of a single quality index we have two of them, running time and final fitness.

Table 1. Ranges of PSO parameters delimiting the search space of the tuning algorithms. During NSGA-II evolution, they are normalized in the range $[0, 1]$ and linearly scaled in the correct range when the PSO instance is created.

Parameter	Range
Population size	$[4, 300]$
Inertia factor (w)	$[-0.5, 2.0]$
c_1	$[-1.0, 4.0]$
c_2	$[-1.0, 4.0]$
Topology	$\{ring, star, global\}$

The function to optimize is a simple instance of the problem presented in Sect. 3. A point cloud representing a wooden mallet is randomly roto-translated and used as target. The same point cloud is used as reference, so it is possible to reach a perfect alignment if PSO is able to find the correct roto-translation.

6   Results

Figure 4 shows the results obtained by EMOPaT. Each point represents the result of a set of parameters on the two objectives. The ones highlighted are the non-dominated ones that approximate the Pareto front and are the ones to be taken into consideration.

Let us have a brief look at these solutions. The main difference that we can observe between parameter sets that converge quickly to a solution with respect to the ones that reach a high final precision is that the former have a smaller population. This confirms the results we obtained on benchmark functions [13]; small populations are good at reaching quickly a good fitness value, but are generally unable to refine it because they are more likely to get stuck into local minima. Nevertheless, good populations are usually smaller than the ones usually suggested by common rules of thumb (e.g. ten times the problem dimension).

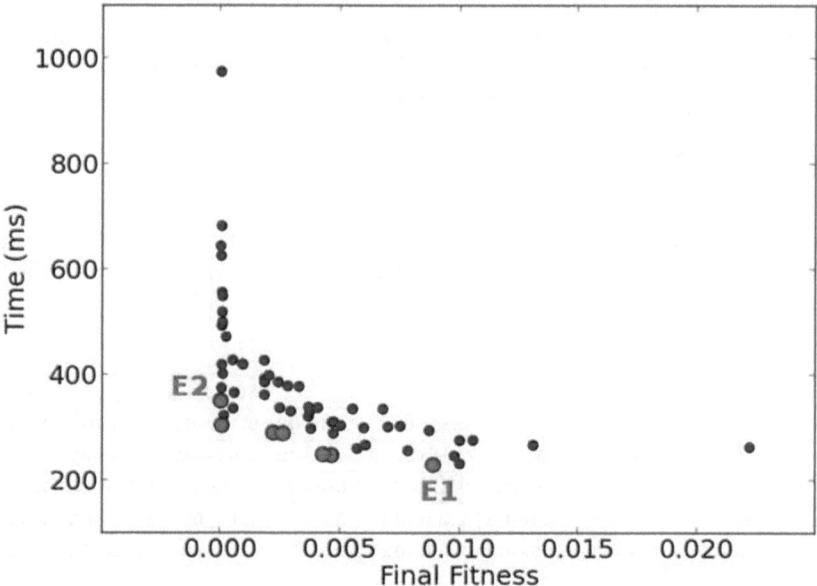

Fig. 4. Results of EMOPaT. Larger, red dots represent the solutions approximating the Pareto front. E_1 and E_2 indicate the "fastest" and the "most precise" solutions, respectively (Color figure online).

We selected the two solutions which lie on opposite ends of the Pareto Front approximation (indicated by E_1 and E_2 in Fig. 4), and tested them more deeply on the point-cloud alignment problem. To prove the correctness of the results, we compared the parameters found by EMOPaT with the ones used in [14]. The latter parameters had been selected as the ones that performed best within a

Table 2. Manually tuned (M_1 and M_2) PSO instances and instances optimized by EMOPaT (E_1 and E_2). The termination criterion is time (0.3, 0.7, 1.3, 2.3 and 3.2 s).

Parameter	M_1	M_2	E_1	E_2
Population size	24	24	15	22
Inertia factor (w)	0.5	0.7	0.6652	0.5944
c_1	1.19	1.8	1.0479	2.1320
c_2	1.19	0.7	0.4271	0.7120
Topology	*Ring*	*Global*	*Global*	*Global*

set of 40 manually-chosen combinations. These parameters, as well as the ones found by EMOPaT are listed in Table 2.

To test them, we performed the same task used in the optimization process using each set of parameters with different time limits, similarly to what we have done in [14]. The time limits we chose were 0.3, 0.7, 1.3, 2.3 and 3.2 s.

Table 3 shows, for each PSO configuration and for each time limit, the average fitness and standard deviation computed over 100 independent repetitions. We performed the Wilcoxon signed-rank test ($p = 0.01$) between results at each time limit to see if there were significant differences; for each time limit, the best-performing PSOs are highlighted in bold.

Table 3. Average and standard deviation of fitness. Each column shows data of a PSO version, each row shows all data obtained at the corresponding time limit. PSO versions that perform statistically better are highlighted in bold.

PSO →	M_1		M_2		E_1		E_2	
Time ↓	Avg	Std	Avg	Std	Avg	Std	Avg	Std
0.3 s	1.62e-01	9.34e-02	1.24e-01	5.89e-02	**7.66e-02**	**9.11e-02**	1.10e-01	1.12e-01
0.7 s	4.37e-02	2.96e-02	3.65e-02	2.41e-02	**2.48e-02**	**2.64e-02**	**2.05e-02**	**1.54e-02**
1.3 s	1.80e-02	1.55e-02	1.41e-02	2.61e-02	1.68e-02	2.01e-02	**9.30e-03**	**1.62e-02**
2.3 s	**7.48e-03**	**7.83e-03**	**7.64e-03**	**1.24e-02**	1.30e-02	1.44e-02	**6.65e-03**	**1.42e-02**
3.2 s	**4.02e-03**	**6.36e-03**	**5.65e-03**	**1.04e-02**	1.29e-02	1.25e-02	**4.53e-03**	**8.62e-03**

These results prove the correctness of our multi-objective approach. The set of parameters E_1, which was the fastest to reach the fitness goal, is in fact the best-performing one after 0.3 s, but it is not able to further improve its results. At the end of evolution, E_2 is comparable to the two manually-selected ones. The main difference between E_2 and the latter is clear if one considers the intermediate time limits, on which E_2 performs better than M_1 and M_2. This is a clear advantage of using EMOPaT. Eventually, many parameter combinations are able to constantly reach a perfect solution if a sufficient amount of time is given. EMOPaT has the ability to find, among all these possible solutions, the ones that are also able to reach a good fitness as fast as possible, because it also optimizes the other objective(s), that are not taken into consideration in a single-objective optimization or are very difficult to account for when performing manual tuning.

The fitness we chose is only a surrogate that reflects the real goal, which is to align the poses of the target and the reference point clouds. In Fig. 5 we present the same results from a different point of view. The plots show the average error in terms of translation (Euclidean distance between centers of gravity of target and reference at the end of optimization) and rotation (angle between the orientations of target and reference). These data allow one to draw the same conclusions and confirm the correctness of the proposed approach.

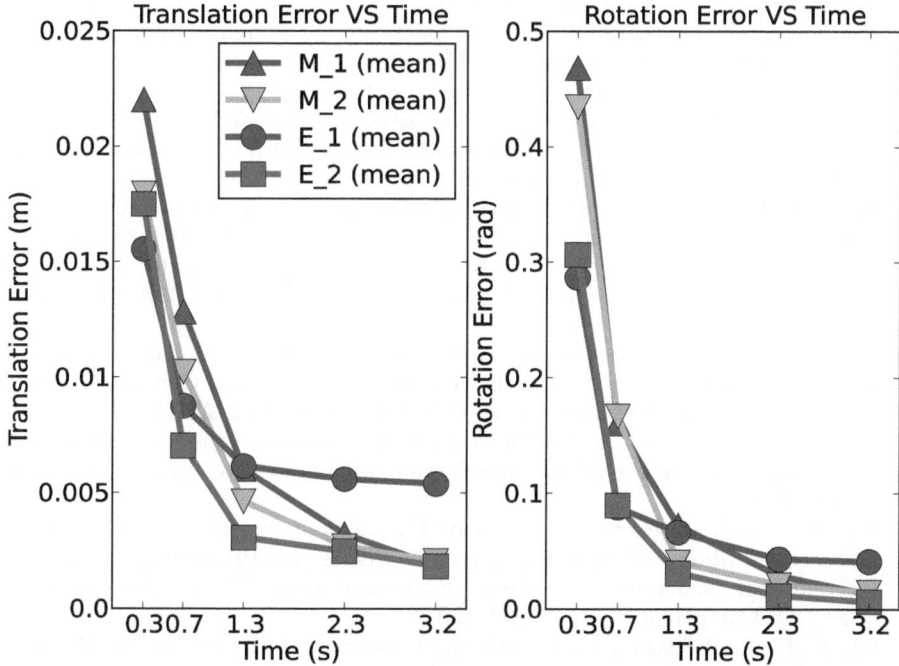

Fig. 5. Results of the four PSOs with different time budgets.

7   Final Remarks

In this paper we proved the ability of multi-objective automatic parameter tuning to optimize PSO's performance on a real-world task, the localization of an object represented as a point cloud.

Using a multi-objective metaheuristic, we were able to obtain a "fast" PSO version that reached good results in a timely manner and an "optimized" PSO which, at convergence, reached results comparable to the manually-tuned solutions, but more quickly, because in the optimization phase this ability was explicitly requested. This proves the potentiality of our multi-objective approach to parameter tuning in real-world optimization tasks, in which the best compromise between conflicting goals such as time and quality of results is usually requested.

References

1. Bartz-Beielstein, T., Lasarczyk, C., Preuss, M.: Sequential parameter optimization. In: IEEE Congress on Evolutionary Computation, vol. 1, pp. 773–780 (2005)
2. Burke, E.K., Hyde, M., Kendall, G., Ochoa, G., Özcan, E.: A classification of hyper-heuristic approaches. In: Gendreau, M., Potvin, J.-Y. (eds.) Handbook of Metaheuristics. International Series in Operations Research & Management Science, vol. 146, pp. 449–468. Springer, Heidelberg (2010)
3. Deb, K., Pratap, A., Agarwal, S., Meyarivan, T.: A fast and elitist multiobjective genetic algorithm: NSGA-II. IEEE Trans. Evol. Comput. **6**(2), 182–197 (2002)
4. Eiben, A.E., Smit, S.K.: Parameter tuning for configuring and analyzing evolutionary algorithms. Swarm Evol. Comput. **1**(1), 19–31 (2011)
5. Kennedy, J., Eberhart, R.: Particle Swarm Optimization. In: Proceedings of IEEE International Conference on Neural Networks. vol. 4, pp. 1942–1948 (1995)
6. Li, H., Shen, T., Huang, X.: Approximately global optimization for robust alignment of generalized shapes. IEEE Trans. Pattern Anal. Mach. Intell. **33**(6), 1116–1131 (2011)
7. López-Ibánez, M., Dubois-Lacoste, J., Stützle, T., Birattari, M.: The irace package, iterated race for automatic algorithm configuration. IRIDIA, Université Libre de Bruxelles, Belgium, Technical report TR/IRIDIA/2011-004 (2011)
8. Makadia, A., Patterson, A., Daniilidis, K.: Fully automatic registration of 3D point clouds. In: Conference on Computer Vision and Pattern Recognition, pp. 1297–1304 (2006)
9. Nashed, Y.S.G., Ugolotti, R., Mesejo, P., Cagnoni, S.: LibCudaOptimize: an open source library of GPU-based metaheuristics. In: Proceedings of the 14th International Conference on Genetic and Evolutionary Computation Conference (GECCO) Companion, pp. 117–124 (2012)
10. Oleari, F., Lodi Rizzini, D., Caselli, S.: A low-cost stereo system for 3D object recognition. In: IEEE International Conference on Intelligent Computer Communication and Processing (ICCP), pp. 127–132 (2013)
11. Rusu, R.B., Blodow, N., Beetz, M.: Fast point feature histograms (FPFH) for 3D registration. In: IEEE International Conference on Robotics and Automation (ICRA), pp. 3212–3217 (2009)
12. Storn, R., Price, K.: Differential evolution - a simple and efficient adaptive scheme for global optimization over continuous spaces. Technical report, International Computer Science Institute (1995)
13. Ugolotti, R., Cagnoni, S.: Analysis of evolutionary algorithms using multi-objective parameter tuning. In: Proceedings of the Genetic and Evolutionary Computation Conference, GECCO, pp. 1343–1350 (2014)
14. Ugolotti, R., Micconi, G., Aleotti, J., Cagnoni, S.: GPU-based point cloud recognition using evolutionary algorithms. In: European Conference on the Applications of Evolutionary Computation, EvoApps (2014)

15. Ugolotti, R., Nashed, Y.S.G., Mesejo, P., Iveković, Š., Mussi, L., Cagnoni, S.: Particle swarm optimization and differential evolution for model-based object detection. Appl. Soft Comput. **13**(6), 3092–3105 (2013)
16. Urfalıoğlu, O., Mikulastik, P.A., Stegmann, I.: Scale invariant robust registration of 3D-point data and a triangle mesh by global optimization. In: Blanc-Talon, J., Philips, W., Popescu, D., Scheunders, P. (eds.) ACIVS 2006. LNCS, vol. 4179, pp. 1059–1070. Springer, Heidelberg (2006)

A Multithreaded Implementation
of the Fish School Search Algorithm

Marcelo Gomes Pereira de Lacerda[1]([✉]) and Fernando Buarque de Lima Neto[1,2]

[1] University of Pernambuco, Recife, Pernambuco, Brazil
{mgpl,fbln}@ecomp.poli.br
[2] Chair for Information Systems and Supply Chain Management,
Westfälische Wilhelms-Universität Münster, 48149 Münster, Germany
buarque@uni-muenster.de

Abstract. This work introduces a multithreaded implementation of the Fish School Search (FSS) algorithm, the Multithreaded Fish School Search (MTFSS). In this new approach, each fish has its behaviour executed within an individual thread, of which creation, execution and death are managed by the runtime environment and the operating system. Five well-known benchmark functions were used in order to evaluate the speed-up of the MTFSS in comparison with the standard FSS and check if there are statistically significant changes in the ability of the new algorithm to find good solutions. The experiments were carried out in a regular personal computer as opposed to expensive set ups and the results showed that the new version of the algorithm is able to achieve interesting growing speed-ups for increasingly higher problem dimensionalities when compared to the standard FSS. This, without losing the ability of the original algorithm of finding good solutions and without any need of more powerful hardware (*e.g.* parallel computers).

Keywords: Computational intelligence · Swarm intelligence · Fish school search · Multithreads

1 Introduction

Optimization tasks are present in many situations where information technology is required. Managers, for example, must take decisions aiming to maximize the companys profit. Racing teams adjust their cars in a way that they will have the best performance given the limits of the machine. These are just some real life examples where optimization tasks are required. Formally, optimization should be understood as a search in which system adjustments according to the utility function are carried out aiming to obtain the best possible outcome (definition extended from Engelbrecht [1]).

Optimization algorithms are computational techniques that search for solutions of problems represented by an objective function. Up to the end of the 1980s, exact optimization algorithms were considered the definitive methods for solving optimization problems. Although significant advances were made on

C. Pizzuti and G. Spezzano (Eds.): WIVACE 2014, CCIS 445, pp. 86–98, 2014.
DOI: 10.1007/978-3-319-12745-3_8

these approaches, it was found that for highly dimensional real-world problems (*e.g.* Supply Chain Network Planning problem [2]), these methods may take an exponentially increasing large amount of time to be solved. Thus, some instances can become intractable.

As an alternative, approximate approaches have been developed in order to face this issue. These techniques do not guarantee the best output possible, but most of time, good enough ones are normally the case.

In this context, nature inspired techniques have been developed. A successful set of these techniques are known as population based algorithms (PBA), due to their characteristics of using a group of artificial entities to collectively and in a coordinated way perform the search.

Swarm Intelligence (SI) can be referred to as the property of any system in which the interaction between very simple components generates complex functional patterns [4]. Within the field of Computational Intelligence, many PBAs present this most interesting behavior. These algorithms form the Swarm Intelligence subfield. Some of the best known algorithms within SI are: Particle Swarm Optimization (PSO) [5], Ant Colony Optimization (ACO) [6], Artificial Bee Colony (ABC) [7], Bacterial Foraging Algorithm (BFA) [8] and Fish School Search (FSS) [9].

The quality of the best solutions found by these algorithms depends on the number of analyzed solutions and, consequently, on the execution time, which means that, in order to acquire better solutions, a longer running time is necessary [3]. Even for some metaheuristics, by increasing the number of dimensions of an optimization problem, one observes an increase of its search spaces complexity. Therefore, a longer running time will be needed for satisfactorily solving these problems. This causes a significant growth in the execution time of the whole optimization process.

Efforts have been made on the creation of parallel versions of SI algorithms. Two main approaches were found in the literature: GPU based parallel algorithms (multicore parallel algorithms), and multithread parallel algorithms. Most of the GPU based parallel algorithms achieved outstanding speed-ups due to the extremely high number of cores available (*e.g.* [10,11]).

Within the multithread parallel algorithms, two approaches were found: cluster based parallel algorithms and single machine parallel algorithms. In comparison to the GPU based approaches, the cluster based parallel algorithms achieved intermediate results, since there are usually a smaller number of available cores when compared to approaches based on GPUs, besides there is a significant adding cost due to the need of communication between processors (*e.g.* [3]). Obviously, the single machine parallel algorithms have been achieved the worst results, in terms of speed-up, among all approaches (*e.g.* [12,13]). However, the main advantage of this approach is that it does not require the acquisition of a complex to run or more expensive system in order to achieve speed-ups. Moreover, the later produces easier coding algorithms than the GPU and cluster based approaches.

Fish School Search (FSS), which was proposed by Bastos Filho and Lima Neto in 2008, is, in its basic version, an unimodal optimization algorithm inspired on the collective behavior of fish schools. The mechanisms of feeding and coordinated movement were used as inspiration to create the collective search mechanism. The main idea is to make the fishes (*i.e.* candidate solutions) to swim toward the direction of the positive gradient in order to gain weight. Collectively, the heavier fishes are more influent in the search process as a whole as the barycenter of the school gradually moves towards better places in the search space. It was firstly designed to run in a single thread.

The first parallel version of the FSS algorithm was proposed by Lins in 2012 [10]. In this work, a GPU based approach (*i.e.* multicore approach) was used in order to speed-up the execution of the FSS algorithm. In the experiments, aiming to evaluate the proposed approach in different architectures, two different machines were used: MacBook Pro with one GPU Nvidia GeForce 320M with 48 cores and one Personal Super Computer (PSC) with 4 GPUs Tesla C2070 with 448 cores each one, all working in parallel. All GPUs used in the experiments are compatible with CUDA architecture. The first one was equipped with an Intel processor with 2 cores and the latter with 4 Intel processors with 4 cores each one. There the authors achieved a speed-up up to 127.9006 with all GPUs working together, besides its performance in terms of fitness of the best solution found was significantly improved in comparison with the FSS algorithm.

This work proposes a single machine multithread version of the Fish School Search algorithm. In this version, there is no need for a more expensive and more complex platform in order to run the referred algorithm. The same machine that was only able to run the standard version of the FSS was able to run the new faster approach, presented here. This new version was specially designed for highly dimensional problems, usually the real-world ones, due to the large execution time needed to optimize such problems.

This paper is organized as follows: on Sect. 2 the standard Fish School Search algorithm is explained; on Sect. 3, the Multithreaded Fish School Search is presented; on Sect. 4, the experiments performed in this work are described; on Sect. 5, the results are showed and discussed; and finally, on Sect. 6, a conclusion about this work is made and all future work is presented.

2   Fish School Search

Fish School Search is inspired by the collective behaviour of natural fish schools. In fish schools, the individuals work collectively as a single organism but do possess some local freedom. This combination accounts for fine as well as greater granularities during their search for food.

In FSS, the success of the search process is represented by the weight of each fish. In other words, the heavier is an individual, the better is its represented solution. The weight of the fish is updated throughout the feeding process. A second means to encode success in FSS is the radius of the school. It is noteworthy to mention that by contracting or expanding the radius of the school, FSS can automatically switches between exploitation and exploration, respectively. The pseudo-code of FSS is provided by Algorithm 1.

P: Fish population;
All individuals are initialized;
while *Stopping condition is not met* **do**
> **foreach** *Fish f in P* **do** run the individual movement;
> **foreach** *Fish f in P* **do** run the feeding process;
> **foreach** *Fish f in P* **do** run the collective instinctive movement;
> Calculate the fish school's barycenter;
> **foreach** *Fish f in P* **do** run the collective volitive movement;

end
Return the best solution found;
Algorithm 1. Pseudo-code of FSS.

2.1 Individual Movement Operator

In the Individual Movement Operator, each fish moves randomly and independently, but always toward the positive gradient. In other words, the fish moves only if the new position is better than the previous one, with regards to the objective function. This movement is described by (1), where $\overrightarrow{x}_i(t+1)$ in the position vector of the individual i, $\overrightarrow{x}_i(t)$ is its old position, $rand(0,1)$ is a random value between 0 and 1 and $step_{ind}(t)$ is the step size on time t. The new step size is calculated through (2), where $step_{ind_{init}}$ and $step_{ind_{final}}$ are the initial and final step sizes and *iterations* is the maximum number of iterations.

$$\overrightarrow{x}_i(t+1) = \overrightarrow{x}_i(t) + rand(0,1)\ step_{ind}(t), \tag{1}$$

$$step_{ind}(t+1) = step_{ind}(t) - \frac{step_{ind_{init}} - step_{ind_{final}}}{iterations}. \tag{2}$$

2.2 Feeding Operator

As mentioned before, the feeding operator is responsible for the weight update of all fishes. This update process is defined by (3), where Δf_i is the fitness variation after the Individual Movement of the fish i, and $max(\Delta f)$ is the maximum fitness variation in the whole population.

$$W_i(t+1) = W_i(t) + \frac{\Delta f_i}{max(\Delta f)}. \tag{3}$$

2.3 Collective Instinctive Movement Operator

The Collective Instinctive Movement Operator is the first collective movement in the algorithm. Every fish performs this movement by adding a vector to its current position, which is calculated according to (4), where N is the population size, $\Delta \overrightarrow{x}_k$ and $\Delta f(\overrightarrow{x}_k)$ are the position variation and the fitness variation of the individual of index in the Individual Movement, being this vector common to all fishes. The final movement is defined by (5).

$$\overrightarrow{I} = \left(\frac{\sum_{k=1}^{N} \Delta \overrightarrow{x}_k \Delta f(\overrightarrow{x}_k)}{\sum_{k=1}^{N} \Delta f(\overrightarrow{x}_k)} \right) \tag{4}$$

$$\overrightarrow{x}_i(t+1) = \overrightarrow{x}_i(t) + \overrightarrow{I}. \tag{5}$$

2.4   Collective Volitive Movement Operator

In this step, the population must contract or expand, using as reference the barycenter of the fish school, which is calculated according to (6), where $W_i(t)$ is the weight of the fish on time t. The total weight of the whole population must be calculated in order to decide if the fish school will contract or expand. If the total weight increased after the last Individual Movement, the school as a whole will contract in order to execute a finer search, which means that the search process has been successful. Otherwise, the population will expand, meaning that the search process is not qualitatively improving. This could be due to a bad region of the search space or the school is trapped in a local minimum (hence, it should try to escape from it). The contraction and expansion processes are defined by (7) and (8), respectively, where $distance(\overrightarrow{x}_i(t), \overrightarrow{B}_j(t))$, is the Euclidian distance between the vectors $\overrightarrow{x}_i(t)$ and $\overrightarrow{B}_j(t))$ and $step_{vol}$ is the volitive step size, which must be defined by the user.

$$\overrightarrow{B}_j(t) = \frac{\sum_{i=1}^{N} \overrightarrow{x}_i(t) W_i(t)}{\sum_{i=1}^{N} W_i(t)}, \tag{6}$$

$$\overrightarrow{x}_i(t+1) = \overrightarrow{x}_i(t) - step_{vol} rand(0,1) \frac{(\overrightarrow{x}_i(t) - \overrightarrow{B}_j(t))}{distance(\overrightarrow{x}_i(t), \overrightarrow{B}_j(t))}, \tag{7}$$

$$\overrightarrow{x}_i(t+1) = \overrightarrow{x}_i(t) + step_{vol} rand(0,1) \frac{(\overrightarrow{x}_i(t) - \overrightarrow{B}_j(t))}{distance(\overrightarrow{x}_i(t), \overrightarrow{B}_j(t))}. \tag{8}$$

3   Multithreaded Fish School Search

The Multithreaded Fish School Search (MTFSS) algorithm is designed to run a single machine, splitting the search process into parallel threads in order to reduce the running time of the optimization task. In this algorithm, the behaviour of each fish in the population is executed in one individual thread. The pseudocode of the algorithm performed by each fish is showed in Algorithm 2.

while *Stopping condition is not met* **do**
 | Perform the individual movement;
 | Barrier1();
 | Perform the feeding process;
 | Barrier2();
 | Perform the collective instinctive movement;
 | Perform the collective volitive movement;
end

Algorithm 2. Algorithm performed by each fish (*i.e.* each thread).

The individual movement is performed exactly like it is made in the standard FSS algorithm. After this operator, the fish calls a sub-routine called *Barrier1*. This sub-routine is showed in Algorithm 3.

static barrier1_counter:Number of fishes that is WAITING on Barrier1();
population_size:Number of fishes in the population;
if *barrier1_counter<population_size* **then**
 | barrier1_counter = barrier1_counter+1;
 | **wait()**;
else
 | Find maximum fitness variation and store in a global variable, which is accessible by all fishes;
 | Calculate $\overrightarrow{I}$ vector for collective instinctive movement;
 | **notifyAll()**;
end

Algorithm 3. Barrier1 sub-routine.

It is important to mention that algorithm was coded in Java programming language version 7. Therefore, the considered *static* modifier used in the pseudo-code of *Barrier1* sub-routine makes the instance of the variable *barrier1_counter* unique for all fishes. Moreover, the **wait()** and **notifyAll()** functions make the fish that calls the sub-routine *Barrier1* enter in WAITING state and wake up all the other fishes, putting them in RUNNABLE state again, respectively. It means that every fish that reaches this sub-routine must wait for the other ones. The last fish to reach this barrier finds the maximum fitness variation, shares this information with the others, calculates the vector $\overrightarrow{I}$ and wakes-up the rest of the population.

After all fishes are in RUNNABLE state again, each one calculates its new weight throughout the feeding operator. After the feeding operator, the *Barrier2* sub-routine is called. This sub-routine is showed in Algorithm 4.

After all fishes are in RUNNABLE state again, the collective instinctive and volitive movements are performed exactly like the standard FSS algorithm, using the vector $\overrightarrow{I}$, the barycentre and fish school total weight calculated by the last fish reach- ing *Barrier2* sub-routine. However, there is a small change in the barycentre calculation, but still important: all fishes positions used for this purpose are the ones right after the execution of the Individual Movement, instead of being the positions after the Insitinctive Movement.

These barriers are intended to keep a minimum synchronism among the fishes. The only change in the algorithms rationale in comparison to the standard version is the use of the positions after the individual movement in the calculation of the barycentre, instead of using the positions after the collective instinctive movement. Results, which are presented in Sect. 5, proved that this change did not produce significant changes in the algorithms ability to find good solutions.

It is important to mention that all threads execution, creation and death are automatically managed by the runtime environment and the operating system, which in itself is another facilitator for the adoption of the multi-threaded version of the FSS algorithm.

> *static* barrier2_counter:Number of fishes that is WAITING on Barrier2();
> population_size:Number of fishes in the population;
> **if** *barrier2_counter<population_size* **then**
> > barrier2_counter = barrier2_counter+1;
> > **wait();**
>
> **else**
> > Calculate weight sum of all fishes;
> > Calculate barycentre, taking into account the position of each fish after the individual movement;
> > **notifyAll();**
>
> **end**

Algorithm 4. Barrier2 sub-routine.

4   Experiments Description

This work aims to compare the running time and the ability of finding good solutions of the MTFSS with the FSS algorithm. Five well known benchmark functions were used for this purpose: Rastrigin, Rosenbrock, Griewank, Ackley and Schwefel 1.2, which are described in [9]. The boundaries of the search space of these functions were set to $[-5.12, 5.12]$, $[-30, 30]$, $[-600, 600]$, $[-32, 32]$, $[-100, 100]$, respectively. The initialization subspaces for each function were set to [2.56,5.12], [15,30], [300,600], [16,32], [50,100]. The initialization subspace defines the region in the search space in which all individuals must be initialized.

Thirty executions were performed for each experiment. For both algorithms, the individual step size and the W_{scale} factor were set to 10 % of the search space length, at the beginning of the optimization task, linearly decreasing to 0.001 %, at the end of the process, and 5000, respectively. The volitive step size was set to twice the individual step size. For the evaluation of the best fitness found, 30 fishes, 30 dimensions for each function and 5000 iterations were used. For running time evaluation, the number of dimensions, individuals and iterations were varied, as shown on Sect. 5, but only Rastrigin function was used.

The execution time were measured in milliseconds for each execution. The speed-up values are the average value of the speed-ups in the thirty executions for each combination between dimensions number and population size.

All experiments were performed in a personal computer equipped with a processor Intel Core i5 650 3.2 GHz (2 physical and 2 virtual cores), 4 GB DDR3 RAM memory. The operational system installed was Ubuntu 13.04 32-bits, with Java version 1.7.0_25 and OpenJDK 7 Runtime Environment IcedTea 2.3.10.

5   Results and Discussion

In this section, all results acquired throughout the experiments are presented.

Figures 1, 2 and 3 show the speed-ups achieved by the MTFSS in comparison to the running time of the FSS algorithm for 100, 1000 and 10000 iterations, respectively. It is possible to see that for 10 and 100 dimensions, the FSS is faster than the MTFSS. However, throughout the approximation of the linear functions that are formed by the points that represents the results of the experiments for 100 and 1000 dimensions, it can be also perceived that the MTFSS is quite faster than the FSS when the problem has approximately more than 203, 309 and 297 dimensions for 100, 1000 and 10000 iterations, respectively. Since there are no results for experiments with the number of dimensions between 100 and 1000, linear functions were used in order to acquire these approximations. The linear functions used for this purpose were: $f(x) = 0.00127(x - 100) + 0.871$, $f(x) = 0.00167(x - 100) + 0.65$ and $f(x) = 0.00165(x - 100) + 0.6746$.

Since the MTFSS was designed to reduce the running time of the FSS algorithm in cases where the standard version is taking too long to complete the

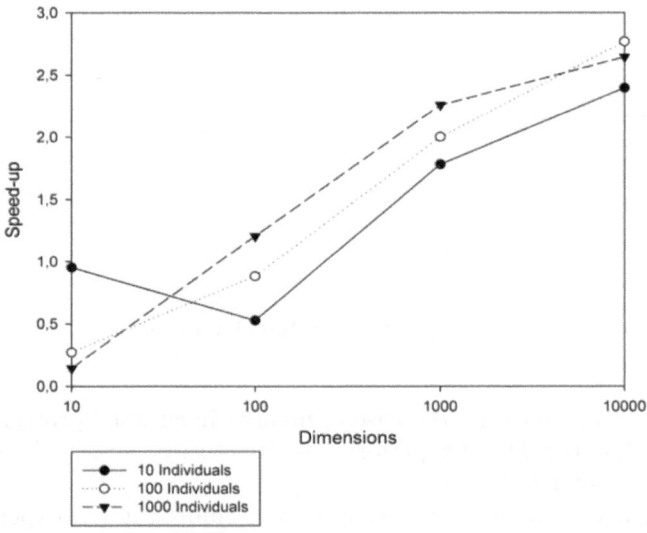

Fig. 1. Speed-up for 100 iterations.

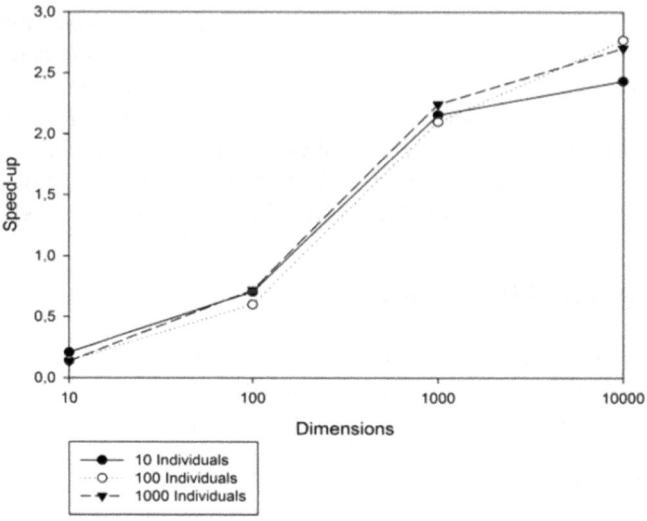

Fig. 2. Speed-up for 1000 iterations.

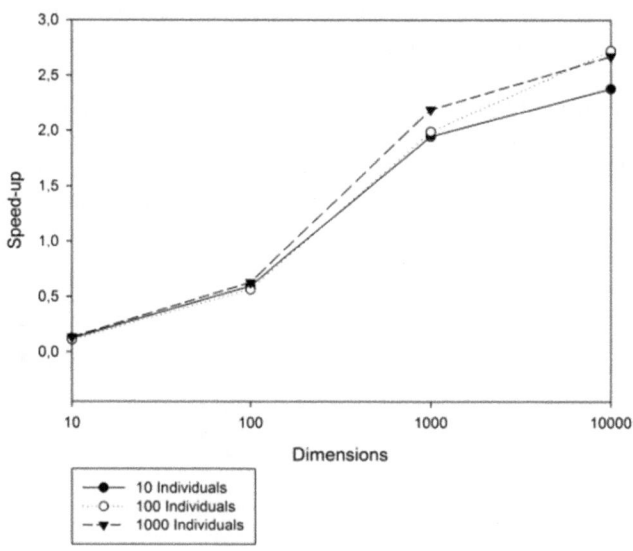

Fig. 3. Speed-up for 10000 iterations.

optimization task, which are the case of highly dimensional problems, it is possible to say that the MTFSS performs satisfactorily those tasks without any technical or economical burden.

It is clear also that the best results were accquired in the experiments with larger dimensions, *e.g.* over 1000 dimensions; continuing to improve even further for 10,000 dimensions. It was not observed significant differences between the experiments with 1000 and 10000 iterations. Even though the experiments with 100 iterations presented some significantly different results from the ones for 10, 100 and 1000 dimensions, for 10000 dimensions the results were quite similar. The three highest speed-ups were achieved in the experiments with 100 individuals and 10000 dimensions for 100, 1000 and 10000 iterations: 2.7695, 2.7668 and 2.7234, respectively.

Figures 4, 5, 6, 7 and 8 shows comparisons between the FSS and MTFSS algorithms in terms of best fitness found. The box-plot graphs represents 30 executions with the setup presented on Sect. 4 in each function. Observing these graphs in conjunction with the Wilcoxon test results that are presented on Table 1, it can be affirmed that there is no significant difference in terms of best fitness found between the algorithms FSS and MTFSS. Therefore, it can be concluded that the MTFSS is able to speed-up the FSS execution up to approximately 3 times, without losing the ability of finding good solutions of the original algorithm.

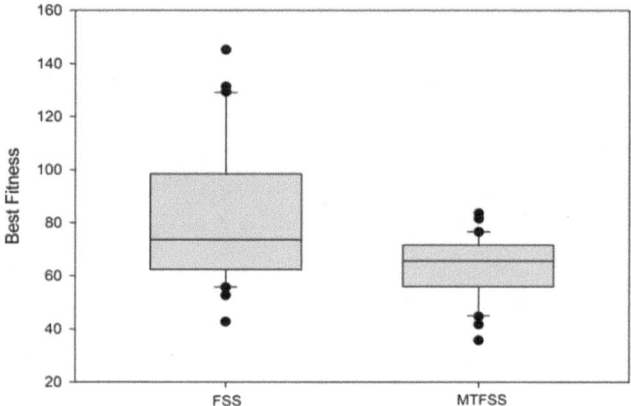

Fig. 4. Best fitness found comparison for Rastrigin.

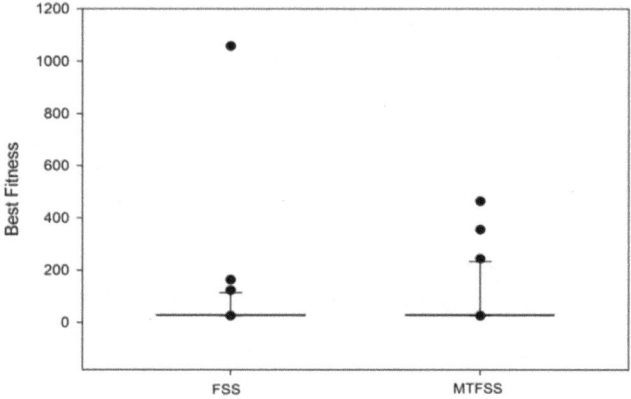

Fig. 5. Best fitness found comparison for Rosenbrock.

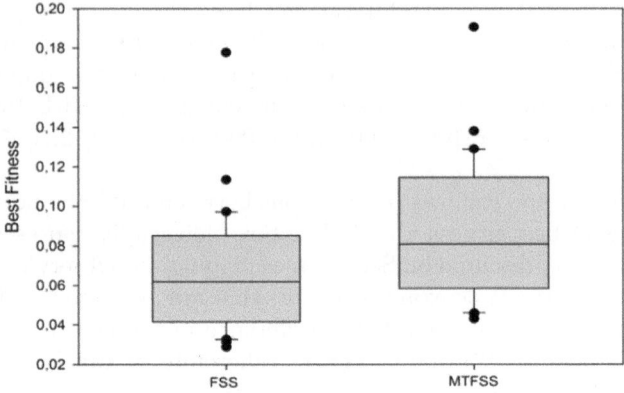

Fig. 6. Best fitness found comparison for Griewank.

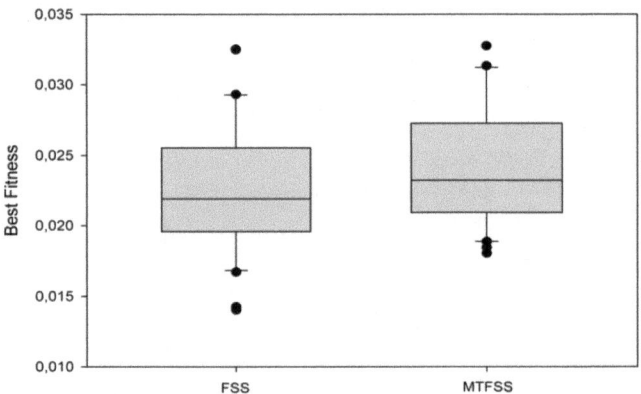

Fig. 7. Best fitness found comparison for Ackley.

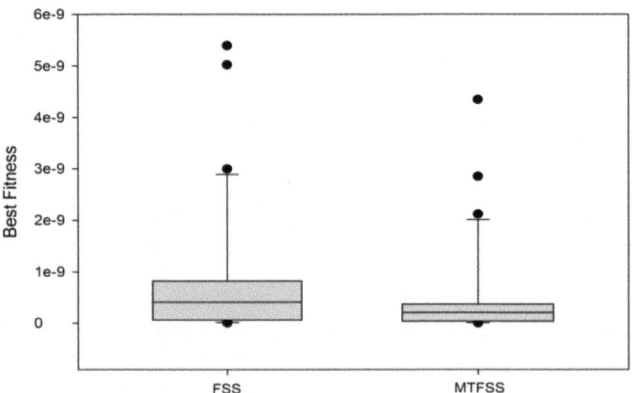

Fig. 8. Best fitness found comparison for Schwefel 1.2.

Table 1. Wilcoxon test results for best fitness found, comparison between MTFSS and FSS. (p = 0.05)

Objective function	FSS
Rastrigin	+
Rosenbrock	X
Griewank	-
Ackley	X
Schwefel 1.2	X

6   Conclusion and Future Work

In this paper, it was presented a multithreaded version of the FSS algorithm, the Multithreaded Fish School Search algorithm (MTFSS). Differently from the pFSS algorithm, which is the first parallel version of the FSS (GPU based implementation), the parallelism of the MTFSS is based on threads, instead of cores, where the behaviour of each fish is run by a single thread. So that, the new approach was designed to run in any personal computer with multithread processing available, without the need for GPUs or even, of a cluster.

The results showed that the algorithm is able to achieve interesting speedups in comparison to the standard FSS, without losing the ability of the original algorithm of finding good solutions.

As future work, the authors of this work intend to test the algorithm in computers with more powerful processors; test the algorithm in different operating systems; and test the algorithm with different programming languages.

On Table 1, '+', '-' and 'X' mean that the MTFSS performed statistically better, worse or equal to the FSS, respectively.

References

1. Engelbrecht, A.P.: Computational Intelligence, An Introduction. Wiley, New Jersey (2007)
2. Pessoa, L.F.A., Horstkemper, D., Braga, D.S., Hellingrath, B., Lacerda, M.G.P., Lima Neto, F.B.: Comparison of optimization techniques for complex supply chain network planning problems. In: Anais do Congresso Nacional de Pesquisa e Ensino em Transporte (ANPET), Belm-Brazil (2013)
3. Bozejko, W., Pempera, J., Smutnicki, C.: Multi-thread parallel metaheuristics for the flow shop problem. In: Artificial Intelligence and Soft Computing, pp. 454–462 (2008)
4. Bonabeau, E., Dorigo, M., Theraulaz, G.: Swarm Intelligence: from Natural to Artificial Systems. Oxford University Press Inc., New York (1999)
5. Kennedy, J., Eberhart, R.: A new optimizer using particle swarm theory. In: International Symposium on Micro Machine and Human Science, pp. 39–43 (1995)
6. Dorigo, M: Optimization, learning and natural algorithms. Ph.D. Thesis Politecnico di Milano (1992)

7. Karaboga, D., Basturk, B.: A powerful and efficient algorithm for numerical function optimization: artificial bee colony (abc) algorithm. Global Optimization, Inc. (2006)
8. Passino, K.M.: Biomimicry of bacterial foraging for distributed optimization and control. IEEE Control Syst. Mag. **22**(3), 52–67 (2002)
9. Filho, C.J.A.B., de Lima Neto, F.B., Lins, A.J.C.C., Nascimento, A.I.S., Lima, M.P.: A novel search algorithm based on fish school behavior. In: IEEE International Conference on Systems, Man and Cybernetics, pp. 2646–2651 (2008)
10. Lins, A.J.C.C.: Paralelizao de Algoritmos de Otimizao baseados em Cardumes atravs de Unidades de Procesamento Grfico. MSc Thesis - University of Pernambuco (2012)
11. Ding, K., Zheng, S., Tan, Y.: A GPU-based parallel fireworks algorithm for optimization. In: Genetic and Evolutionary Computation Conference, pp. 9–16 (1999)
12. Bacanin, N., Tuba, M., Brajevic, I.: Performance of object-oriented software system for im- proved artificial bee colony optimization. Int. J. Math. Comput. Simul. **5**(2), 154–162 (2011)
13. Tuba, M., Bacanin, N., Stanarevic N.: Multithreaded implementation and performance of a modified artificial fish swarm algorithm for unconstrained optimization. Int. J. Math. Comput. Simul., 215–222 (1999)

Evolutionary Applications to Cellular Automata Models for Volcano Risk Mitigation

Giuseppe Filippone[1], Roberto Parise[1], Davide Spataro[1],
Donato D'Ambrosio[1,2], Rocco Rongo[1,2], and William Spataro[1,2(✉)]

[1] Department of Mathematics and Computer Science, University of Calabria,
Via Pietro Bucci, 87036 Rende, Italy
spataro@unical.it
[2] CNR-IRPI, Sezione di Cosenza, Rende, Italy

Abstract. A GPGPU accelerated evolutionary computation-based decision support system for defining and optimizing volcanic hazard mitigation interventions is proposed. Specifically, the new Cellular Automata numerical model SCIARA-fv3 for simulating lava flows at Mt Etna (Italy) and Parallel Genetic Algorithms (PGA) have been applied for optimizing protective measures construction by morphological evolution. A case study is considered, where PGA are applied for the optimization of the position, orientation and extension of earth barriers built to protect a touristic facility located near the summit of Mt. Etna (Italy) volcano which was interested by the 2001 lava eruption. The methodology has produced extremely positive results and, in our opinion, can be applied within a broader risk assessment framework, having immediate and far reaching implications both in land use and civil defense planning.

Keywords: Evolutionary computation · Parallel genetic algorithms · Decision support system · Cellular automata · Morphological evolution

1 Introduction

In the modelling and simulation field, Complex Cellular Automata (CCA) can represent a valid methodology to model numerous complex non-linear phenomena [1], such as lava and debris flows. CCA are an extension of classical Cellular Automata (CA) [2], developed for overcoming some of the limitations affecting conventional CA frames such as the modelling of large scale complex phenomena. Due to their particular nature and local dynamics, CCA are very powerful in dealing with complex boundaries, incorporating microscopic interactions and easy parallelization of algorithms.

In lava risk mitigation, the building of artificial barriers [3,4] is fundamental for controlling and slowing down the destructive effects of flows in volcanic areas. Nevertheless, the proper positioning of protective measures in a specific area may depend on many factors (viscosity of the magma, output rates, volume erupted, steepness of the slope, topography, economic costs). As a consequence, in this context, one of the major scientific challenges for volcanologists is to provide efficient and effective solutions.

© Springer International Publishing Switzerland 2014
C. Pizzuti and G. Spezzano (Eds.): WIVACE 2014, CCIS 445, pp. 99–112, 2014.
DOI: 10.1007/978-3-319-12745-3_9

Morphological Evolution (ME) is a recent development within the field of engineering design, by which evolutionary computation techniques are used to tackle complex design projects. This branch of evolutionary computation is also known as evolutionary design and integrates concepts from evolutionary algorithms, engineering and complex systems to solve engineering design problems [5]. Morphological evolution has also been largely explored in evolutionary robotics, for instance in the design of imaginary 3D robotics bodies [6].

Genetic Algorithms (GAs) [7] are general-purpose iterative search algorithms inspired by natural selection and genetics and have been applied several times in the past for optimizing CCA models (e.g., [8,9]). This work describes the application of morphological evolution by Parallel Genetic Algorithms (PGAs) for optimizing earth barriers construction to divert a case study lava flow, occurred in Mt. Etna in 2001. The GA fitness function, adopted for evaluating the "goodness" of the protective works deviating lava flow scenarios generated by the new SCIARA-fv3 CA lava flow model [10], has implied a massive use of the numerical simulator, consisting in thousands of concurrent simulations for every GA generation. Therefore, a GPGPU (General Purpose computation with Graphic Processor Units) library was developed to accelerate the GA execution.

After a brief description of the new SCIARA-fv3 CA model adopted in experiments (Sect. 2), the main characteristics of the implemented evolutionary algorithm and carried out experiments, together with reference to emergent behaviors, are presented in Sect. 3. The developed Web user interface for interactive visualization of results are described in Sect. 4. Eventually, Sect. 5 concludes the paper with final comments and future works.

2   Complex Cellular Automata and the SCIARA-fv3 Lava Flow Model

As previously stated, CCA represent an extension of the classical homogeneous CA, particularly useful for the modeling of spatially extended systems. Formally, a CCA is a 7-tuple:

$$A = < R, X, Q, P, \tau, L, \gamma >$$

where R, X, Q and τ are as in the homogeneous CA definition, respectively defining the CA space, cell neighborhood, set of states and the deterministic transition function applied to each cell, simultaneously and at discrete steps. However, in the CCA frame, the set Q of state of the cell is decomposed in substates, $Q_1, Q_2, \ldots, Q_r$, each one representing a particular feature of the phenomenon to be modelled. The overall state of the cell is thus obtained as the Cartesian product of the considered substates: $Q = Q_1 \times Q_2 \times \ldots \times Q_r$. A set of parameters, $P = p_1, p_2, \ldots, p_p$, is furthermore considered. These allow to calibrate the model for reproducing different dynamics. As the set of states is split in substates, also the transition function, τ, is split in elementary processes, $\tau_1, \tau_2, \ldots, \tau_s$, each one describing a particular aspect that rules the dynamics of the considered phenomenon. Eventually, $L \in R$ is a subset of the cellular space that is subject to external influences, as specified by the supplementary function γ. External influences are introduced in order to model features that are not easy to be described in terms of "local interactions".

2.1   The SCIARA-fv3 Lava Flow Model

SCIARA-fv3 is the latest release of the SCIARA family of Complex Cellular Automata Models for simulating lava flows. As its predecessor, SCIARA-fv2 [11], it is based on a Bingham-like rheology. While more specific details can be found in [10], we briefly describe the model in the following. In formal terms, SCIARA-fv3 is defined as:

$$SCIARA - fv3 =< R, X, Q, P, \tau, L, \gamma >$$

where:

1. R is the cellular space, i.e. the set of square cells covering the bi-dimensional finite region where the phenomenon evolves;
2. X is the pattern of cells belonging to the adjacent eight-cell Moore neighborhood that influence the cell state change;
3. $Q = Q_z \times Q_h \times Q_T \times Q_{\overrightarrow{p}} \times Q_f^9 \times Q_{\overrightarrow{v_f}}^9$ is the finite set of states, considered as Cartesian product of substates. Their meanings are: cell altitude a.s.l., cell lava thickness, cell lava temperature, momentum (both x and y components), lava thickness outflows (from the central cell toward the adjacent cells) and flows velocities (both x and y components), respectively;
4. $P = w, t_0, P_T, P_d, P_{hc}, \delta, \rho, \epsilon, \sigma, c_v$ is the finite set of parameters (invariant in time and space), whose meaning can be found in [10];
5. $\tau : Q^9 \rightarrow Q$ is the cell deterministic transition function; it is split in "elementary processes", which are described in the following sections;
6. $L \subseteq R$ specifies the lava source cells (i.e. craters);
7. $\gamma : Q_h \times \mathbb{N} \rightarrow Q_h$ specifies the emitted lava thickness from the source cells at each step $k \in \mathbb{N}$.

2.2   Elementary Process τ_1: Lava Flows Computation

The elementary process τ_1 computes lava outflows and their velocities, formally defined as:

$$\tau_1 : Q_z^9 \times Q_h^9 \times Q_{\overrightarrow{p}} \rightarrow Q_f^9 \times Q_{\overrightarrow{v_f}}^9$$

Lava flows are computed by a two-step process: the first computes the CA clock, t, i.e. the physical time corresponding to a CA computational step, while the second the effective lava outflows, $h_{(0,i)}$, their velocities $v_{f_{(0,i)}}$ and displacements $s_{(0,i)}$ ($i = 0, 1, ..., 8$). The elementary process τ_1 is thus executed two times, the first one in "time evaluation mode", the second in "flow computing mode". Both modes compute the so called "minimizing outflows", $\phi_{(0,i)}$, i.e. those which minimize the unbalance conditions within the neighborhood, besides their final velocities and displacements. In "time evaluation mode", t is preliminary set to a large value, $t_{\max}$, and the computed displacement, $s_{(0,i)}$, is compared with the maximum allowed value, $d_{(0,i)}$, which is set to the distance between the central cell and the neighbor that receives the flow. In case of over-displacement, the time t must be opportunely reduced in order to avoid the overflow condition (i.e., a flow going beyond the cell neighbourhood). In case no over-displacement are obtained, t remains unchanged. Eventually, in "flow computing mode", effective lava outflows, $h_{(0,i)}$, are computed by adopting the CA clock obtained in "time evaluation mode", by guarantying no overflow condition.

Outflows Computation. In "flow computing mode", minimizing outflows, $\phi_{(0,i)}$, are re-computed by considering the new CA clock $\bar{t}$. Subsequently, lava outflows, $h_{(0,i)}$, are computed proportionally to the displacement, by simply multiplying the minimizing outflow by the ratio between the actual displacement, $s_{(0,i)}$, and the maximum allowed, the cell side w:

$$h_{(0,i)} = \phi_{(0,i)} \frac{s_{(0,i)}}{w}$$

2.3   Elementary Process τ_2: Updating of Mass and Momentum

The elementary process updates lava thickness and momentum. It is formally defined as:

$$\tau_2 : Q_f^9 \times Q_{\overrightarrow{v_f}}^9 \to Q_h \times Q_{\overrightarrow{p}}$$

Once the outflows $h_{(0,i)}$ are known for each cell $c \in R$, the new lava thickness inside the cell can be obtained by considering the mass balance between inflows and outflows:

$$h_{(0)} = \sum_{i=0}^{9}(h_{(i,0)} - h_{(0,i)})$$

Moreover, also the new value for the momentum can be updated by accumulating the contributions given by the inflows:

$$\overrightarrow{p}_{(0)} = \sum_{i=0}^{9} h_{(i,0)}\overrightarrow{v_f}_{(i,0)}$$

2.4   Elementary Process τ_3: Temperature Variation and Lava Solidification

$$\tau_3 : Q_f^9 \times Q_T^9 \to Q_T \times Q_h$$

As in the elementary process τ_1, a two step process determines the new cell lava temperature. In the first one, a temperature $\overline{T}$ is obtained as weighted average of residual lava inside the cell and lava inflows from neighboring ones. A further step updates the calculated temperature $\overline{T}$ by considering thermal energy loss due to lava surface radiation:

$$T = \frac{\overline{T}}{\sqrt[3]{1 + \frac{3\overline{T}^3 \epsilon \sigma \bar{t} \delta}{\rho c_v w^2 h}}}$$

where ϵ, σ, $\bar{t}$, δ, ρ, c_v, w and h are the lava emissivity, the Stephan-Boltzmann constant, the CA clock, the cooling parameter, the lava density, the specific heat, the cell side and the lava thickness, respectively. When the lava temperature drops below the threshold T_{sol}, lava solidifies. Consequently, the cell altitude increases by an amount equal to lava thickness and new lava thickness is set to zero. For more details on all above models specifications, please refer to [10].

3   Morphological Evolution of Protective Works

Genetic Algorithms (GAs) [7] are general-purpose iterative search algorithms inspired by natural selection and genetics, and have been extensively applied in many scientific fields (e.g., [12–14]). GAs simulate the evolution of a population of candidate solutions, called phenotypes, to a specific problem by favouring the reproduction of the best individuals. Phenotypes are codified by genotypes whose elements are called genes. To determine the best possible solution of a given problem, members of the initial population are evaluated by means of a "fitness function", determining the individuals "adaptivity" value. Best individuals are chosen by means of a "selection" operator and reproduced by applying random "genetic" operators to form a new population of offspring. Typical genetic operators are "crossover" and "mutation": they represent a metaphor of sexual reproduction and of genetic mutation, respectively. The overall sequence of fitness assignment, selection, crossover, and mutation is repeated over many generations (i.e. the GA iterations) producing new populations of individuals.

In this work, GAs were adopted in conjunction with the SCIARA-fv3 CA model for the morphological evolution of protective works to control lava flows in the Rifugio Sapienza area, which was indeed interested by the 2001 Nicolosi case study and object of a set of real mitigation interventions construnction [15].

3.1   Parallel Genetic Algorithm Definition

The CA numerical model finite set of states was extended by introducing two substates defined as:

$$Z \subseteq R \tag{1}$$

where Z is the set of cells of the cellular automaton that specifies the Safety Zone, which delimitates the area that has to be protected by the lava flow and

$$P \subseteq R, P \cap Z = \oslash \tag{2}$$

where P is the set of CA cells that identifies the Protection Measures Zone identifying the area in which the protection works are to be located.

The Protection work $W = B_1, B_2, \ldots, B_n$ was represented as a set of barriers, where every barrier $B_i = N_{i1}, N_{i2}$ is composed by a pair of nodes

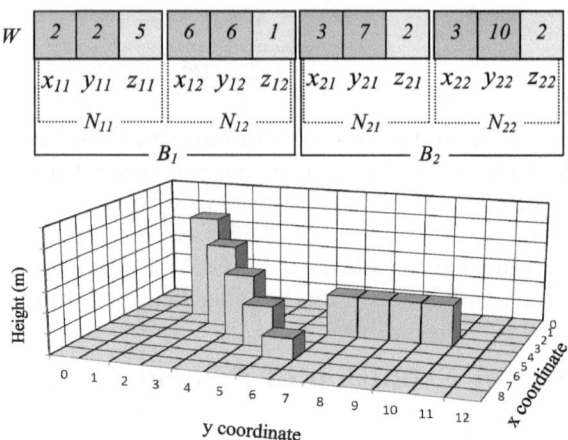

Fig. 1. Example of barriers encoding into a GA genotype. The height of the intermediate points of each barrier is obtained by connecting the work protections extremes through a linear function.

$N_{ij} = x_{ij}, y_{ij}, z_{ij}$, where x_{ij}, y_{ij} are the CA coordinates for the generic node j of the barrier i, and z_{ij} the height (in m). The solutions were encoded into a GA genotype as integer values (Fig. 1) and a population of 100 individuals, randomly generated inside the Protection Measures Zone, was considered.

Two different fitness functions were considered to suitably evaluate the goodness of a given solution: f_1, based on the areal comparison between the simulated event and the Safety Zone (in terms of affected area) and f_2, which considers the total volume of the protection works in order to reduce intervention costs and environmental impact. More formally, the f_1 objective function is defined as:

$$f_1 = \frac{\mu(S \cap Z)}{\mu(Z)} \qquad (3)$$

where S and Z respectively identify the areal extent of the simulated lava event and the Safety Zone area, with $\mu(S \cap Z)$ being the measure of their intersection. The function f_1, assumes values within the range $[0, 1]$ where 0 occurs when the simulated event and Safety Zone Area are completely disjointed (best possible simulation) and 1 occurs when simulated event and Safety Zone Area perfectly overlap (worst possible simulation).

The f_2 objective function is defined as:

$$f_2 = \frac{\sum_{i=1}^{|W|} p_c \cdot d(B_i) \cdot h(B_i)}{V_{max}} \qquad (4)$$

where $d(B_i)$ and $h(B_i)$ represent the length (in meters) and the average height of the i-th barrier, respectively. The parameter p_c is the CA cell side and $V_{max} \in R$ is a threshold parameter (i.e., the maximum building volume) given by experts, for the function normalization. Since the barriers are composed of two nodes, the function can be written as:

$$f_2 = \frac{\sum_{i=1}^{|W|} p_c \cdot d(N_{i1}, N_{i2}) \cdot \bar{h}(N_{i1}, N_{i2})}{V_{max}} \tag{5}$$

where $\bar{h}(N_{i1}, N_{i2}) = \frac{|z_{i1} + z_{i2}|}{2}$ is considered as the average height value between two different nodes and $d(N_{i1}, N_{i2}) = \sqrt{(x_{i1} - x_{i2})^2 + (y_{i1} - y_{i2})^2}$ identifies the Euclidean distance between them. The final fitness function f_2 is thus:

$$f_2 = \frac{\sum_{i=1}^{|W|} p_c \cdot \sqrt{(x_{i1} - x_{i2})^2 + (y_{i1} - y_{i2})^2} \cdot \frac{|z_{i1} + z_{i2}|}{2}}{V_{max}} \tag{6}$$

The function f_2 assumes values within the range $[0, 1]$: it is nearly 0 when the work protection is the cheapest possible, 1 otherwise.

For the genotype fitness evaluation, a composite (aggregate) function f_3 was also introduced as follows:

$$f_3 = f_1 \cdot \omega_1 + f_2 \cdot \omega_2 \tag{7}$$

where $\omega_1, \omega_2 \in R$ and $(\omega_1 + \omega_2) = 1$, represent weight parameters associated to f_1 and f_2. Several different values where tested and the considered ones in this work chosen on the basis of trial and error techniques. The goal for the GA is to find a solution that minimizes the considered objective function $f_3 \in [0, 1]$.

In order to classify each genotype in the population, at every generation run, the algorithm executes the following steps:

1. CA cells elevation a.s.l. are increased/decreased in height on the basis of the genotype decoding (i.e., the barrier cells). In addition, the determination of the cells inside the segment between the work protection extremes and f_2 subsequently are computed.
2. A SCIARA-fv3 simulation is performed (about 40000 calculation steps) and the impact of the lava thickness on Z area (f_1 computation) is evaluated.
3. f_3 is computed and individuals are sorted according to their fitness.

The adopted GA is a rank based and elitist model, as at each step only the best genotypes generate off-spring. The 20 individuals which have the highest fitness generate five off-spring each and the $20 \times 5 = 100$ offspring constitute the next generation. After the rank based selection, the mutation operator is applied with the exception of the first 5 individuals.

The complete list of GA characteristics and parameters is reported in Table 1. Each gene mutation probability depends on its representation: p_{mc} for genes corresponding to coordinates value and p_{mh} viceversa. Therefore, if during the mutation process, a coordinate gene is chosen to be modified, the new value will depend on the parameters x_{max} and y_{max} which represent the cell radius within the node, the position of which can vary. The interval $[h_{min}, h_{max}]$ is the range within which the values of height nodes are allowed to vary (Fig. 2).

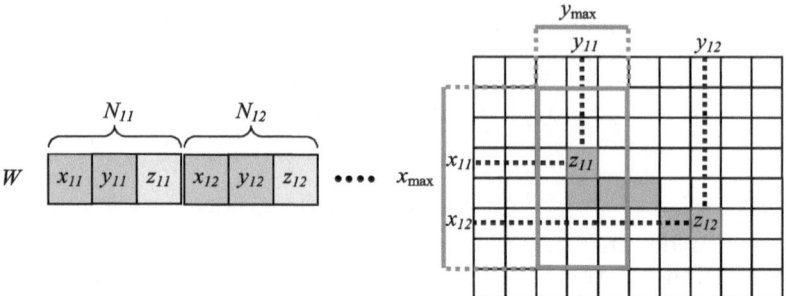

Fig. 2. Graphical representation of the genotype mutation phase. Each gene, representing a CA coordinate, can vary within a variation radius $[x_{max}, y_{max}]$.

This strategy ensures the possibility for the GA to provide, as output, either protective barriers or ditches.

To ensure a better exploration of the search space and to avoid a fast convergence of solutions to local optima, a n point crossover operator has been also introduced. Two parent individuals are randomly chosen from the mating pool and two different cutting points for each parents are selected. After the selection portions chosen in the genotype, they are exchanged. The crossover operator is applied according to a prefixed probability, p_c, for each sub-solution encoded in the genotype.

3.2   Experiments and Results

The fitness evaluation of a GA individual consists in an entire CA simulation, followed by a comparison of the obtained result with the actual case study. This phase may require several seconds, or even several hours: for example, on a 2-Quadcore Intel Xeon E5472, 3.00 GHz CPU such evaluation requires approximately 10 min, as at least 40,000 CA steps are required for a simulation. For instance, if the GA population is composed of 100 individuals, the time required to run one seed test (100 generation steps) exceeds 69 days. Moreover, the GA execution can grow, depending on both the extent of the considered area and the number of different tests to run.

As a consequence, a *CPU/GPU* library was developed to accelerate the GA running. Specifically, a "Master-Slave" model was adopted in which the Host-CPU (Master) executes the GA steps (selection, population replacement, mutation and crossover), while GPU cores (slaves) evaluate the individuals fitness (i.e., a complete SCIARA-fv3 simulation). Please refer to [16–18] for more details on the different GPGPU parallel implementations which are adopted for speeding up the GA running.

By considering the Nicolosi lava flow event (barriers uphill from Sapienza Zone), ten parallel GA runs (based on different random seeds) of 100 generation steps were carried out, each one with a different initial population. The elapsed time achieved for the ten GA runs was less than nine hours of computation

Table 1. List of parameters of the adopted GA

GA parameters	Specification	Value
g_l	Genotype's length	6
p_s	Population size	100
n_g	Number of generations	100
p_{mc}	Coord. gene mutation probability	0.4
x_{max}	Gene x position variation radius	10
y_{max}	Gene y position variation radius	10
p_{mh}	Height gene mutation probability	0.45
h_{min}	height min variation range	-5
h_{max}	height max variation range	10
p_c	crossover probability	0.05
c_{h+}	Cost to build	1
c_{h-}	Cost to dig	1
ω_{f1}	f_1 weight parameter	0.95
ω_{f2}	f_2 weight parameter	0.05

Table 2. Properties of the best barrier evolved by GA run

Barrier	Length (m)	Height (m)	Base width (m)	Volume (m^3)	Inclination (degrees)
$[210, 94, 3]$ $[236, 124, 12]$	397	7.5	10	24750	131

on a 10 multi-GPU GTX 680 GPU Kepler Devices Cluster (note that the same experiment, on a sequential machine, would had lasted more than seven months). Furthermore, during the running, a new 3D WebGL visualization system (discussed in Sect. 4) was developed, making the model fully portable and allowing the interactive visualization and analysis phases of the results.

For this experiment, only solutions with two nodes were considered ($|W| = 1$), while Z and P were chosen as in Fig. 5. The cardinality of W (Protection work) and the gene values in which they are allowed to vary (depending of Z area), define the search space S_r for the GA:

$$S_r = \{[P_{x_{min}}, P_{x_{max}}] \times [P_{y_{min}}, P_{y_{max}}] \times [(h_{min} \cdot n_g), (h_{max} \cdot n_g)]\}^{2|W|} \quad (8)$$

The temporal evolution of the f_3 fitness is graphically reported in Fig. 3(a), in terms of average results over the ten considered experiments. GA experiment parameters values are also listed in Table 1. The related CA simulation, obtained by adopting the best individual is shown in Fig. 5.

The study, though preliminary, has produced quite satisfying results. Among different best individuals generated by the GA for each seed test, the best one (Table 2) consists of a barrier with an average height of 7.5 m and 397 m in

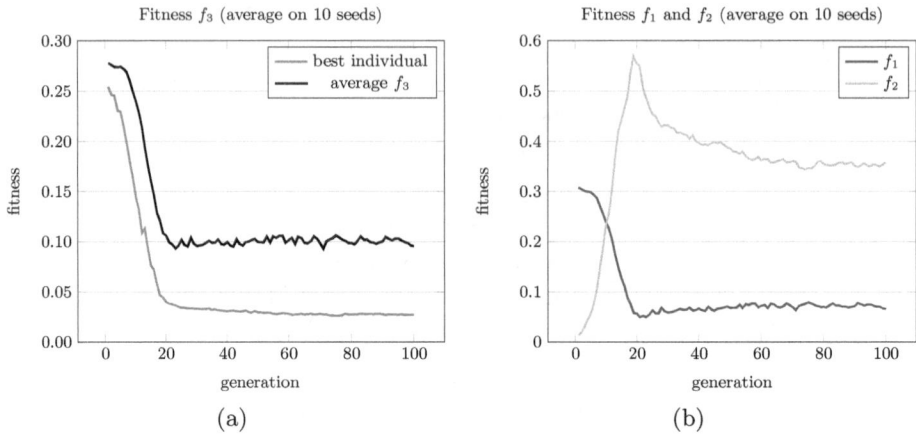

Fig. 3. Temporal evolution of composite f_3 fitness of best individual (in black) and of average fitness of whole population (in gray) (a). Temporal evolution of average fitness f_1 (in red) and f_2 (in green) of whole population (b). Fitness values were obtained as an average of 10 GA runs, adopting different seeds for generation of random numbers (Color figure online).

length with an inclination angle of 131° with respect to the direction of the lava flow. The barrier (cf. Table 2) completely deviates the flow avoiding that the lava reaches the inhabited and building facilities areas. The best solution provided by GA (Fig. 5) in this work is approximately five times more efficient (in term of total m^3 volume used to keep safe tha safety areas) respect to the one applied in the real case [15], consisting of thirteen earthen barriers.

3.3　Considerations on the GA Dynamics and Emergent Behaviors

In the executed GA experiments, individuals with high fitness evolved rapidly, even if the initial population was randomly generated and the search space was quite large (Eq. 8). By analyzing several individuals evolved in ten different GA executions, similar solutions were observed. This behavior is due to the presence of problem constraints (e.g. morphology, lava vent, emission rate, Z and P areas) that lead the GA to search in a "region" of the solution space characterized by a so called "local optimum". In particular, f_1 reaches the minimum value (0) around the twentieth GA generation and the remaining 80 runs are used by GA for the f_2 optimization (cf. Fig. 3 (b)). In any case, the evolutionary process has shown, in accordance with the opinion of the scientific community [3,19], the ineffectiveness of barriers placed perpendicular to the lava flow direction despite diagonally oriented solutions (130° − 160°).

　　Furthermore, a systematic exploitation of morphological characteristics by GA, during the evolutionary process, has emerged. To better investigate such GA emergence behaviour, a study of nodes distribution was conducted (Fig. 4(a)). By considering the solutions provided by GA, each node was classified on the

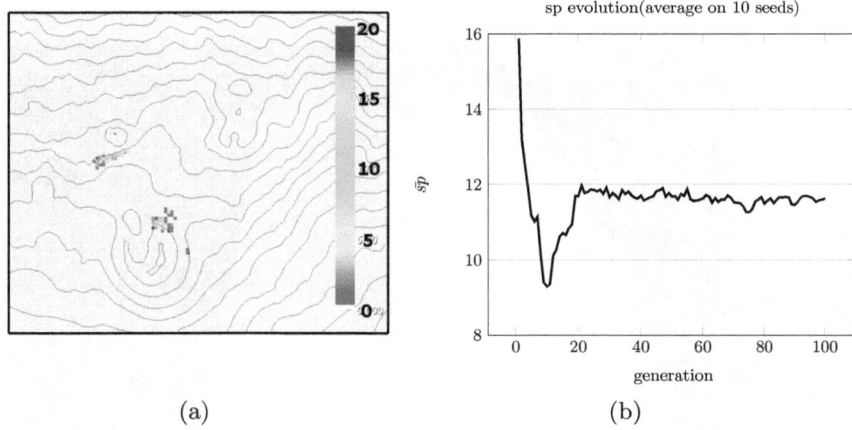

(a)   (b)

Fig. 4. (a) Nodes distribution of the best 100 solutions generated by the GA. Scale values indicates occurrence of nodes. (b) Temporal evolution of average slope proximity values for the best individuals.

slope proximity calculation, as an average of altitude differences between node neighborhood cells (with radius 10 cells) and the central cell. The function that assigns to each generic node j a *slope proximity* value is defined as:

$$sp_j = \frac{\sum_{i=1}^{|X|} \bar{z}_i - \bar{z}_0}{|X|} \tag{9}$$

where X is the set of cells that identifies the neighborhood of j and $\bar{z}_i \in Q_z$ is the topographics altitude (index 0 represents the central cell). As shown in Fig. 4(b), starting from the tenth GA generation, the evolutionary process has shown an increase in slope proximity values. Therefore, after the f_1 optimization (cf. Fig. 3(b)), in order to minimize f_2, there is a specific evolutionary temporal phase (i.e., up to the 25th generation) where the algorithm generates solutions that are located in the proximity of elevated slopes.

4   SciaraWii: The SCIARA-fv3 Web User Interface

The simulation results were visualized in real-time by means of a 3D interactive visualization system based on WebGL, a cross-platform application program interface used to create 3D computer graphics in Web browsers. SCIARAWii, a Web 2.0 application, controls the simulation while the SCIARA-fv3 model runs server-side. The application is based on HTML5 and JavaScript, which permits its full portability. The client is able to control the basic SCIARA-fv3 simulation functionalities thanks to asynchronous callbacks to the server.

SCIARAWii was implemented by means GWT where the interaction between the user interface and the SCIARA-fv3 computational model is performed by a set of client-side services which are implemented on the server. Multi-client

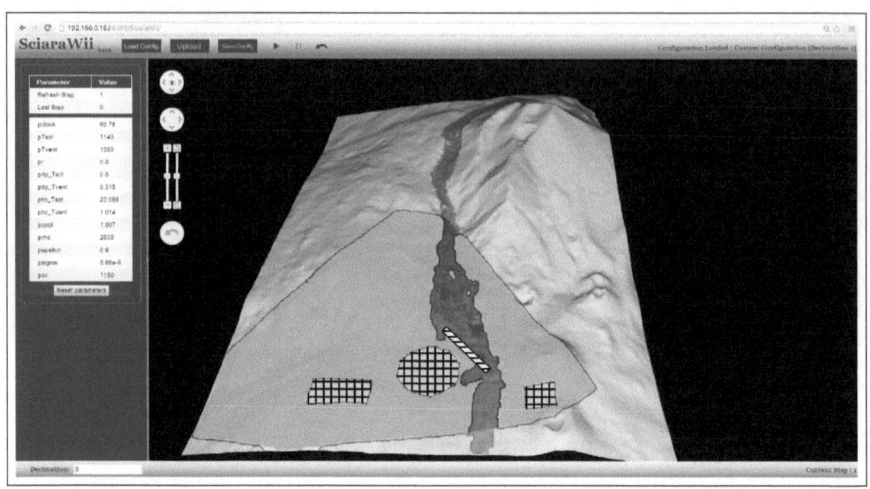

Fig. 5. A screenshot of the Web user interface for SCIARA-fv3 showing 3D simulation of 2001 Nicolosi lava flow adopting the GA best solution of the best solution. As seen, the devised barrier (dashed line area) completely diverts the lava flow from the considered Safety Areas (gridded areas).

connections are also possible: whenever a user logs in, an asynchronous request is sent to the server in order to establish a connection. Here, a servlet binds the client to an individual connection-handler, which allows multiple unambiguous communications through HTTP requests and responses.

The SCIARAWii system architecture is the same as [20]. The computational model, SCIARA-fv3, is implemented on the server in C++ (for efficiency reasons) as a static library. A dynamic-link library (DLL) receives requests by Java Native Interface (JNI) methods and provides simulation data to the application server. Data is therefore sent to the client via HTTP and stored into the Web browser cache memory. SCIARA-fv3 parameters are displayed in GUI controls (in which they can also be modified), while simulation data such as the topographic surface or the simulated lava flow, are visualized by means of the 3D WebGL rendering engine, which runs on a HTML5 `<Canvas>`. Whenever the simulation produces a lava flow, it is displayed over the surface and its dynamical behavior can be observed. All the client-server communications are managed by means of asynchronous JavaScript calls, which are able to provide the same usability level of desktop applications. Figure 5 shows a screenshot of SCIARAWii.

In order to stress the system's reliability, SCIARAWii was tested on a Local Area Network by only considering laptops, with one acting as a Web server and a maximum of 4 as remote clients accessing simultaneously the former. The level of usability of the GUI resulted more than satisfactory, mainly thanks to the asynchronous communications between client and server. Also the 3D visualization system resulted to be surprisingly efficient, especially if compared with that of the first release of SCIARAWii, by making it comparable to standard desktop applications in terms of both efficiency and usability.

5   Conclusions and Future Works

This paper has presented an evolutionary approach for devising protective measures to divert lava flows. Starting from the adoption of the latest release of the SCIARA-fv3 Cellular Automata lava model and adoption of Parallel Genetic Algorithms, a library was developed for executing a large number of concurrent lava simulations using GPGPU and a new WebGL-based 3D visualization system implemented for the real-time result analysis.

First observations of the GA results permitted to conjecture the presence of a local optima in the search space, probably due to problem constraints. To better investigate GA dynamic characteristics, a study of nodes distribution was also conducted and a systematic exploitation of morphological characteristics by GA during the evolutionary process emerged. Nevertheless, PGAs experiments, carried out by considering the Nicolosi case-study, demonstrated that artificial barriers, placed at suitable positions and orientations, can successfully change the direction of lava flow in order to protect predefined point of interests.

The study has produced extremely positive results and simulations have demonstrated that GAs can represent a valid tool to determine protection works construction in order to mitigate the lava flows risk. Future work will consider the investigation of solutions consisting of multiple protective interventions and the SCIARAWii visualization system graphical enrichment to fully exploit the capability of the underlying computational model.

Acknowledgments. This work was partially funded by the European Commission – European Social Fund and by the Regione Calabria (Italy). Authors gratefully acknowledge the support of NVIDIA Corporation for this research.

References

1. Di Gregorio, S., Serra, R.: An empirical method for modelling and simulating some complex macroscopic phenomena by cellular automata. Future Gener. Comput. Syst. **16**(2–3), 259–271 (1999)
2. Von Neumann, J.: Theory Self-reproducing Automata. University of Illinois Press, Champaign (1966)
3. Barberi, F., Brondi, F., Carapezza, M., Cavarra, L., Murgia, C.: Earthen barriers to control lava flows in the 2001 eruption of Mt. Etna. J. Volcanol. Geoth. Res. **123**, 231–243 (2003)
4. Colombrita, R.: Methodology for the construction of earth barriers to divert lava flows: the Mt. Etna 1983 eruption. Bull. Volcanol. **47**(4), 1009–1038 (1984)
5. Bentley, P.: An introduction to evolutionary design by computers (chap. 1). In: Bentley, P.J. (ed.) Evolutionary Design by Computers, pp. 1–73. Morgan Kaufman, San Francisco (1999)

6. Sims, K.: Evolving 3D morphology and behavior by competition. In: Proceedings of Artificial Life IV, pp. 28–39. MIT Press (1994)
7. Holland, J.H.: Adaptation in Natural and Artificial Systems: An Introductory Analysis with Applications to Biology, Control, and Artificial Intelligence. The MIT Press, Cambridge (1992)
8. D'Ambrosio, D., Spataro, W.: Parallel evolutionary modelling of geological processes. J. Parallel Comput. **33**(3), 186–212 (2007)
9. D'Ambrosio, D., Rongo, R., Spataro, W., Trunfio, G.A.: Optimizing cellular automata through a meta-model assisted memetic algorithm. In: Coello, C.A.C., Cutello, V., Deb, K., Forrest, S., Nicosia, G., Pavone, M. (eds.) PPSN 2012, Part II. LNCS, vol. 7492, pp. 317–326. Springer, Heidelberg (2012)
10. D'Ambrosio, D., Spataro, W., Parise, R., Rongo, R., Filippone, G., Spataro, D., Iovine, G., Marocco, D.: Lava flow modeling by the sciara-fv3 parallel numerical code. In: 22nd Euromicro International Conference on Parallel, Distributed, and Network-Based Processing, pp. 330–338 (2014)
11. Radu, V.: Application. In: Radu, V. (ed.) Stochastic Modeling of Thermal Fatigue Crack Growth. ACM, vol. 1, pp. 63–70. Springer, Heidelberg (2015)
12. Hinton, G.E., Nowlan, S.J.: How learning can guide evolution. Complex Syst. **1**, 495–502 (1987)
13. Peters, J.F.: Topology of digital images: basic ingredients. In: Peters, J.F. (ed.) Topology of Digital Images. ISRL, vol. 63, pp. 1–76. Springer, Heidelberg (2014)
14. D'Ambrosio, D., Rongo, R., Spataro, W., Trunfio, G.A.: Meta-model assisted evolutionary optimization of cellular automata: an application to the SCIARA model. In: Wyrzykowski, R., Dongarra, J., Karczewski, K., Waśniewski, J. (eds.) PPAM 2011, Part II. LNCS, vol. 7204, pp. 533–542. Springer, Heidelberg (2012)
15. Barberi, F., Carapezza, M.L.: The control of lava flows at Mt. Etna. In: Bonaccorso, A., Calvari, S., Coltelli, M., Del Negro, C., Falsaperla, S. (eds.) Mt. Etna: Volcano Laboratory, 357th edn, p. 369. American Geophysical Union, Washington, D.C. (2004)
16. Blecic, I., Cecchini, A., Trunfio, G.: Cellular automata simulation of urban dynamics through GPGPU. J. Supercomput. **65**, 614–629 (2013)
17. D'Ambrosio, D., Filippone, G., Marocco, D., Rongo, R., Spataro, W.: Efficient application of GPGPU for lava flow hazard mapping. J. Supercomput. **65**(2), 630–644 (2013)
18. Di Gregorio, S., Filippone, G., Spataro, W., Trunfio, G.A.: Accelerating wildfire susceptibility mapping through GPGPU. J. Parallel Distrib. Comput. **73**(8), 1183–1194 (2013)
19. Fujita, E., Hidaka, M., Goto, A., Umino, S.: Simulations of measures to control lava flows. Bull. Volcanol. **71**, 401–408 (2009)
20. Parise, R., D'Ambrosio, D., Spingola, G., Filippone, G., Rongo, R., Trunfio, G.A., Spataro, W.: Swii2, a HTML5/WebGL application for cellular automata debris flows simulation. In: Sirakoulis, G.C., Bandini, S. (eds.) ACRI 2012. LNCS, vol. 7495, pp. 444–453. Springer, Heidelberg (2012)

On RAF Sets and Autocatalytic Cycles in Random Reaction Networks

Alessandro Filisetti[1]([✉]), Marco Villani[1,2], Chiara Damiani[3], Alex Graudenzi[3], Andrea Roli[4], Wim Hordijk[5], and Roberto Serra[1,2]

[1] European Centre for Living Technology,
University Ca' Foscari of Venice, Venice, Italy
`alessandro.filisetti@unive.it`
[2] Department of Physics, Informatics and Mathematics,
University of Modena and Reggio Emilia, Modena, Italy
[3] Department of Informatics, Systems and Communication,
University of Milano Bicocca, Milan, Italy
[4] Department of Computer Science and Engineering (DISI),
University of Bologna, Bologna, Italy
[5] SmartAnalytiX.com, Lausanne, Switzerland

Abstract. The emergence of autocatalytic sets of molecules seems to have played an important role in the origin of life context. Although the possibility to reproduce this emergence in laboratory has received considerable attention, this is still far from being achieved.

In order to unravel some key properties enabling the emergence of structures potentially able to sustain their own existence and growth, in this work we investigate the probability to observe them in ensembles of random catalytic reaction networks characterized by different structural properties.

From the point of view of network topology, an autocatalytic set have been defined either in term of strongly connected components (SCCs) or as reflexively autocatalytic and food-generated sets (RAFs).

We observe that the average level of catalysis differently affects the probability to observe a SCC or a RAF, highlighting the existence of a region where the former can be observed, whereas the latter cannot. This parameter also affects the composition of the RAF, which can be further characterized into linear structures, autocatalysis or SCCs.

Interestingly, we show that the different network topology (uniform as opposed to power-law catalysis systems) does not have a significantly divergent impact on SCCs and RAFs appearance, whereas the proportion between cleavages and condensations seems instead to play a role.

A major factor that limits the probability of RAF appearance and that may explain some of the difficulties encountered in laboratory seems to be the presence of molecules which can accumulate without being substrate or catalyst of any reaction.

1 Introduction

The dynamics of sets of interacting molecular species, in those cases where new molecular types can be created by the interactions among some of the existing

© Springer International Publishing Switzerland 2014
C. Pizzuti and G. Spezzano (Eds.): WIVACE 2014, CCIS 445, pp. 113–126, 2014.
DOI: 10.1007/978-3-319-12745-3_10

ones, pose formidable problems. This study is of the outmost importance in researches on the origin of life (OoL) but its applications might be interesting also within laboratories or industrial applications.

When only few new molecular types can be generated, the complications are manageable. The actual interesting case is the one where many types might appear. In order to study the generic features of this kind of systems, the use of random molecular species has been introduced [10–12,14], in particular in the context of the so-called binary polymer model [9,21,22] where all the species are linear strings made out of a binary alphabet. Since catalysis plays a prominent role in biological processes, it is often also assumed that only catalyzed reactions take place at an appreciable rate. In the spirit of searching for generic properties, randomness is also extensively used to determine catalysts: every molecular species has a certain fixed probability to catalyze a certain reaction.

In this way it is possible to represent the reactions as a directed graph, where there is an edge from a molecular species to all those species whose production reactions are catalyzed by that species. To complete the picture, it is worth stressing that only two types of reactions are allowed, condensation (which creates a new string by concatenating two previous ones) and cleavage (which generates two strings by separating two parts of a preexisting one). The model is described in more details in Sect. 2.

The graph described above links catalysts to the products of the reactions they catalyze; it can therefore be considered as a *catalyst → product* graph. The products may in turn catalyze other reactions. It has been often suggested that the presence of cycles (similar to those that are found in present-day biological systems [1,6–10,12,20,29]) plays a key role in the dynamics of these systems, since cycles of catalysts give rise to a collectively autocatalytic system. To be precise, we will call *SCC* a Strongly Connected Component [5] of the network described above, i.e. a subset of the entire network where each node is reachable (directly or indirectly) from every other node in the subset. A *SCC* is therefore composed of one or more molecular types where the formation of each member is catalyzed by at least another member of the *SCC*.

While some models (e.g. [19]) suppose that the presence of a catalyst is sufficient for a reaction to occur, in the binary polymer model considered here, in order for cleavage or condensation reactions to occur, the substrates must be present as well. These substrates may belong to what is called the food set F, which involves for example those molecules of the feed those are supposed to be freely available or to be continuously supplied from outside; these molecules might or might not have catalytic properties. Substrates that are not in the food set may be present too, i.e. molecular species that take part in reactions and are produced and consumed by the system itself.

The importance of the substrates is properly captured by the notion of Reflexive Autocatalytic and Food generated (RAF) sets, that is a set of molecular types and reactions where each member can be generated by other members of the set through a series of catalyzed reactions, starting from F [16–18]. Note that the identification of this structure needs knowledge about all the chemical species (substrates, products, catalysts) involved in the reaction scheme: a more detailed description can be found in Sect. 2.

SCCs and RAFs are indeed important tools to analyze and understand the structure of sets of interacting molecules: note that an SCC may not be a RAF, given that substrates are not taken into account, and that a RAF may also not be an SCC, given that cyclic structures are not strictly required (a RAF may turn out not to have cycles if at least one of the species of the food set is also a catalyst).

Here below we analyze how frequent SCCs and RAFs are as a function of the average level of catalysis $\langle c \rangle$ (the average number of reactions catalyzed by each chemical species). Note that the behavior of the model turns out to be trivial when $\langle c \rangle$ is very small (no SCCs and no RAFs are observed) or very high (the system is full of both). Instead, it is interesting to consider an intermediate region. An outcome of the present work is that there exists a region of $\langle c \rangle$, similar results have observed in [17] observing both RA and RAF sets, that we term the gap region, where SCCs can be observed, but RAFs are not (these sentences are to be interpreted in a statistical sense, see the results in Sect. 3 for the details). Therefore, in this region, the very existence of cycles in the catalyst → product networks does not imply that they actually affect in a sensible way the main network properties.

In the present work we will investigate protocell systems, compartments containing sets of molecules able to collectively self-replicate, able to undergo fission and proliferate [26,34]. Several protocell architectures have been proposed [4,13,15,23,24,27,30,32,33], most of them identify the compartment with a lipid vesicle that may spontaneously fission under suitable circumstances. In the specific, we are here interested on a very simple protocell model, where a semi-permeable membrane embrace a small physical space where all the key reactions take place in the aqueous phase, and it allows the passage of small molecules only [31].

In the present work we investigate sets of autocatalytic reactions where a constant concentration of some small chemical species is guaranteed: this situation can be maintained by chemo-physical systems as protocells (compartments containing sets of molecules able to collectively self-replicate, able to undergo fission and proliferate [26,31,34]—an interesting system for both Origin of Life and technological themes) or CSTRs (Continuous-flow Stirred-Tank Reactor [28]).

Thus, the topological structures having the highest probabilities to emerge in such systems, by taking a purely graph theoretical approach, will be investigated, as in previous papers by Kauffman, Steel and Hordijk [17,18], and the existence of the gap region will be shown. It is however tempting to speculate about the behavior of a truly dynamical model in this region.

We have already shown in our previous works [10–12,14] that, for $\langle c \rangle$ values close to the threshold for SCCs appearance, those that are observed tend to be fragile, some of their links refer to very rare reactions and do not really affect the behavior of the system [10,12]. We guess that this behavior is related to the absence of RAFs, and that by exploring the region close to the second threshold (i.e. the one for RAFs) one should be able to find more productive networks. Moreover, we noted that (within the size of the systems we explored) the thresholds of SCCs and RAFs appearance in uniform and power-law catalysis systems are very similar. Interestingly, RAF systems at different average levels of catalysis are composed of different mixtures of autocatalysis, SCCs and linear structures. Different proportions of cleavages and condensations lead to different probabilities of RAFs appearance, and the presence of a large fraction of reactions producing long molecules (condensations ligating long chemical species) could inhibit the RAFs appearance as we will be discussed in Sect. 3. So, the RAF presence is influenced in different ways by different characteristics of the chemical system.

In Sect. 3 the main results are presented, while the discussion of further work is deferred to the final Sect. 4.

2   Description of the Model and Network Topologies

A detailed description of the theoretical model can be found in Kauffman [21,22] and in Filisetti et al. [10,12]. In the following only the main features, useful for the scope of this work, will be summarized. Random catalytic reaction networks are composed of a set of species $S = (s_1, s_2, \cdots, s_n)$. In particular, species are concatenation of characters from left to right taken from a finite alphabet, in this work a binary alphabet $A = \{1; 0\}$. Thus, the cardinality $|S|$ of S is $|S| = 2^{M+1} - 2$, where M is the maximum length of the system. In accordance with the original version of the model [21], each $s_i \in S$ can take part to two types of reaction only: condensation, where two species are concatenated to create a longer species (e.g. $100 + 10 \rightarrow 10010$), and cleavage, where a species is divided in two shorter species (e.g. $1001 \rightarrow 10 + 01$); s_i can be either a substrate or a product according to the position within the reaction scheme. Each species has an independent probability $p_{s_i} = 1/|S|$ to be selected as a substrate for a reaction, hence in case of condensation two species will be selected while in case of cleavage just one species will be selected. The overall number of conceivable reactions $\hat{R}$, i.e., all the reaction schemes allowed by the combination of all the possible $|S|$ molecular species, varies according to the constructional method adopted; in this work we typically adopt the method described in the following.[1]

Reactions are created so that the product of the reaction cannot be longer than M, hence the number of conceivable reactions is equal to:

$$\hat{R} = 2 \cdot \sum_{i=2}^{M} 2^i \cdot (i - 1) \tag{1}$$

where the multiplicative term 2 indicates that each reaction scheme is in principle valid both for cleavage reactions and condensation reactions.[2] Not all the

---

[1] The only exception is presented in the final part of the results section.

[2] Forward and backward reactions, during the creation of the reactions graph, are in principle handled as two separated reactions.

reactions belonging to $\hat{R}$ are catalyzed. So, it is possible to compute an average level of catalysis $\langle c \rangle = R/|S|$ which denotes the average number of reactions catalyzed by each single species.[3] Therefore, each catalysis can be represented as a pair (r_k, s_i) where $r_k \in \hat{R}$ stands for the selected reaction and $s_i \in S$ behaves as a catalyst for r_k.

The procedure of catalysis assignation leads to different catalytic reaction network topologies. In this work, two kinds of topologies will be assessed, based on uniform and preferential assignation. In the former case, each $s_i \in S$ has the same probability $p_k^i = 1/|S|$ to be a catalyst for a whatever reaction k, while in the latter case p_k^i is determined by the chemical characteristics of the involved species. In real systems, researchers observed that very few species can catalyze many reactions, whereas many species catalyze only one reaction, or none. This situation can be represented by means of a catalysis distribution having the shape of a power law, our second topology.[4]

The entire set $\mathcal{C} = (r_k, s_j)$ of catalysis form what we call "artificial chemistry", hence different random instantiations of $\mathcal{C}$ lead to different random catalytic reaction networks, i.e., different "artificial chemistries"; in Fig. 1 an example of a complete reaction graph is depicted while in Fig. 2 the catalyst $\rightarrow$ product representation of the graph shown in Fig. 1 is reported.

The structural role of backward reactions is evaluated as well. Thus, artificial chemistries imposing the presence of both forward and reverse reactions will be created.[5] It is worthwhile to notice that introducing reverse reactions leads to a double number of reactions catalyzed, so in case of reverse reactions the number of reactions to catalyze is divided by 2, $\hat{R} \rightarrow \hat{R}/2$.

Analysis will be carried out measuring the emergence of reflexively autocatalytic sets (RAF sets) on the complete graph $CRN = (S, R, \mathcal{C})$, where $\mathcal{C}$ represents catalysis, and investigating the strongly connected components emerging in the graph representation considering catalysts and products only, hence showing the catalytic activity of the system without taking care of the presence and the nature of the substrates. In Fig. 1 two examples of RAF sets and two SCCs are shown. In particular, the group of molecular species and reactions belonging

---

[3] Since different M will be assessed, in this work we prefer to adopt the average level of catalysis $\langle c \rangle$, as in [16], instead of the standard reaction probability p, as we adopted in our previous works [10,12], and Kauffman [21] and others [2,25] in important works on this topic. Nevertheless, according to $|S|$, it is always possible to move from $\langle c \rangle$ to p and *vice-versa*.

[4] We obtain the power law distribution by slightly modifying the algorithm proposed by Barabási and Albert [3]. We increase p_k^i as a function of the already catalyzed reactions, i.e., the probability to catalyze a reaction is weighted with the number of reactions already catalyzed so that $p_k^i = \#r_i / \sum_{z=1}^{|S|} \#r_z$, where $\#r_i$ and $\#r_z$ indicate the number of reactions already catalyzed by the i-th and the z-th species respectively. In such a way, we obtain a power-law distribution in the number of reactions each chemical species can catalyze.

[5] Since the analysis are static and only statistic structural properties of the networks will be assessed, the choice on which reaction is the direct one and which is the reverse one is only implementative and does not affect the analysis.

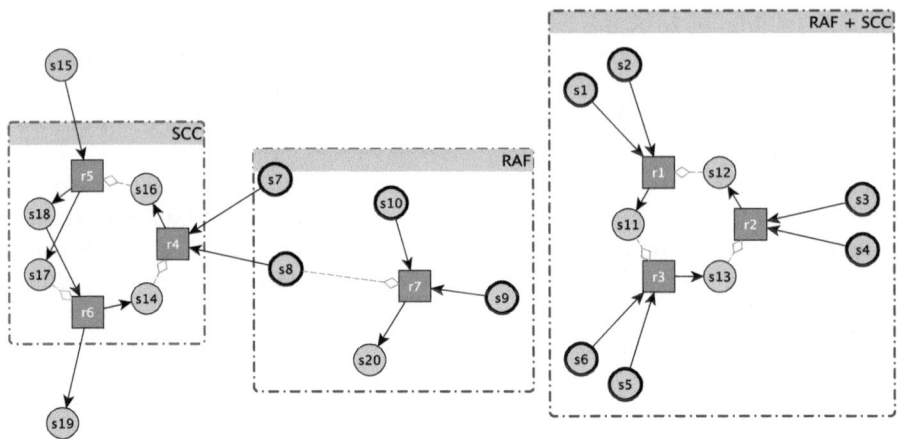

Fig. 1. The picture shows an example of representation of a catalytic reaction network by means of a complete reaction graph, taking into account both molecular species and reactions. Circles stand for molecular species and in particular bold circles represent species belonging to the food set F. Squares depict reactions, straight arrows indicate the participation of a species to the reaction as substrate (edge points to the reaction) or product (edge starts from the reaction). Catalysis are represented by dotted gray arrows. For the sake of comprehension, both molecular species and reactions forming RAFs and SCCs are grouped together.

to the group named "$RAF + SCC$" form a SCC and a RAF at the same time. On the opposite, the other two reaction structures are, from left to right, a SCC only and a RAF set only.

3   Results

As discussed above, we investigate here catalytic reaction networks, where only small molecular species have a constant concentration (this situation can be maintained by protocells [31], or CSTR systems [12]). In the following, scenarios where the length of the buffered species (the "foodset" of the from which RAF structures are built) ranges from 2 ($|S| = 6$) to 4 ($|S| = 62$) will be considered.

We are interested now in analyzing the role of different scenarios on both the presence and the internal structure of RAFs within protocells. The assessed scenarios involve different distributions of the chemicals' catalyzing capabilities (a uniform distribution, and a slightly more realistic power-law distribution, which allows the presence of a little number of highly versatile catalysts), the presence or absence of backward reactions, and the possibility for long chemical species to be not suitable for participating in any reaction, leading in such a way to the accumulation of useless chemical species (for example, because lack of solubility or folding difficulties).

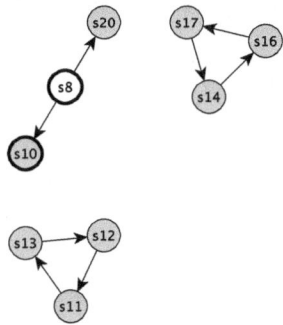

Fig. 2. The picture shows the catalyst $\rightarrow$ product representation of the complete graph $\mathcal{C}$ shown in Fig. 2. In this representation just the catalytic activity of the reaction network is depicted, and no information about the substrate of the reactions are present. Circles stand for molecular species and in particular bold circles represent species belonging to the food set F. In this case straight arrows indicate the catalysis of a particular molecular species by means of its catalyst. By means of this representation, the two SCCs are clearly visible and they are formed from $s_1 1, s_1 2, s_1 3$ and $s_1 4, s_1 6, s_1 7$ respectively.

3.1 Strongly Connected Components and RAFs

The first observation regards the presence of SCCs and RAFs within the reaction graphs in accordance with a foodset composed of species up to length 2 only, i.e., when the membrane allows the transfer of chemicals having maximum length 2. Figure 3 shows the fraction of network instances showing at least 1 SCC (1 RAF), by varying $\langle c \rangle$ and M. It can be observed that the SCCs and RAFs transition zones significantly differ (the higher the maximum length of the system, the higher the transition slope): the vast region between 1.0 and 2.5 levels of catalysis is therefore rich of SCC structures unable to self-sustain because the failed fulfilling the closure condition.[6]

The showed case regards scale-free topologies without backward reactions (obviously, in these experiments we maintain constant the total number of reactions). It is worth mentioning that the bias induced on the system topology by the pairing of forward-backward reactions do not change the positions of the transition zones nor their slopes (data not shown here), as long as condensation and cleavages are equally present - we will discuss the effect of uneven presence of condensation and cleavages in the next paragraph. Remarkably, almost

---

[6] It is important to notice that in [17] the transition is found at 1.25. Such a difference is basically related to a different way to count forward and reverse reactions. Hordijk [17] consider both forward and backward reactions as a unique reaction whereas in our work, though it is clear that they account for the same reaction scheme, we consider them as two different reactions.

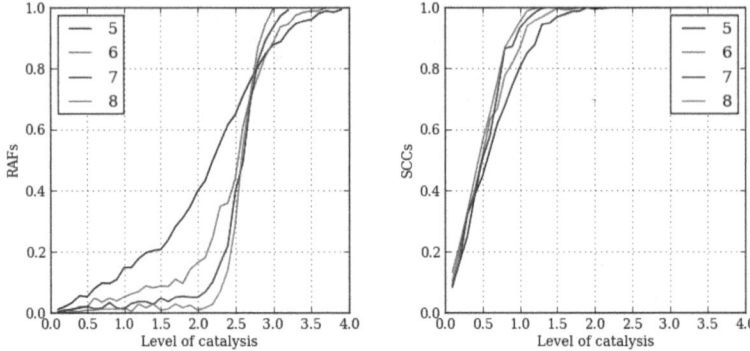

Fig. 3. The fraction of simulations showing at least 1 RAF (left) and 1 SCC (right), by varying $\langle c \rangle$ and M in networks with forward reactions only. F contains all the species up to length 2. On the x-axis the average level of catalysis $\langle c \rangle$ is represented while on the y-axis the fraction of network instances (out of 1000 networks for each $\langle c \rangle$) is depicted.

identical behaviors hold for random topologies having the same number of chemical species and reactions (data not shown, and [17]).[7]

3.2   The Inner RAF Structure

The aforementioned results are derived from systems with protocell membranes that allow the diffusion of chemical species up to length 2 ("BUF_2" scenario). Since F does not show any catalytic activity, in that case all RAFs are necessarily composed of at least one SCC. On the contrary, if we extend the membrane semipermeability to longer species, some of these objects may be able to catalyze some chemical reactions: this is a crucial change, because the formation of RAFs composed of linear structures (on the catalyst $\rightarrow$ products graph) turn out to be possible, thus catalytic roots are buffered ("BUF_3" scenario).

This change in F composition has an apparently huge effect on the presence of RAF, which appears also at low level of catalysis (Fig. 4); note however that the difference among chemistries with different maximum lengths M ("MaxLen M" systems) consists mainly in lower and lower probabilities of choosing the buffered species as root of linear structures on the catalyst $\rightarrow$ products graph (a more interesting use of the "MaxLen M" scenarios is presented in the next paragraph). Indeed, the presence of the 2.5 threshold can be grasped on the

---

[7] In [17], the authors show the so-called RA sets—which are reflexively autocatalytic but not necessarily food-generated—whose transition and nature correspond to that of SCC, where the transitions happen on the same zones of scale-free topologies in case of non catalytic activity of the foodset molecular species, i.e., "BUF_2" scenario in this work.

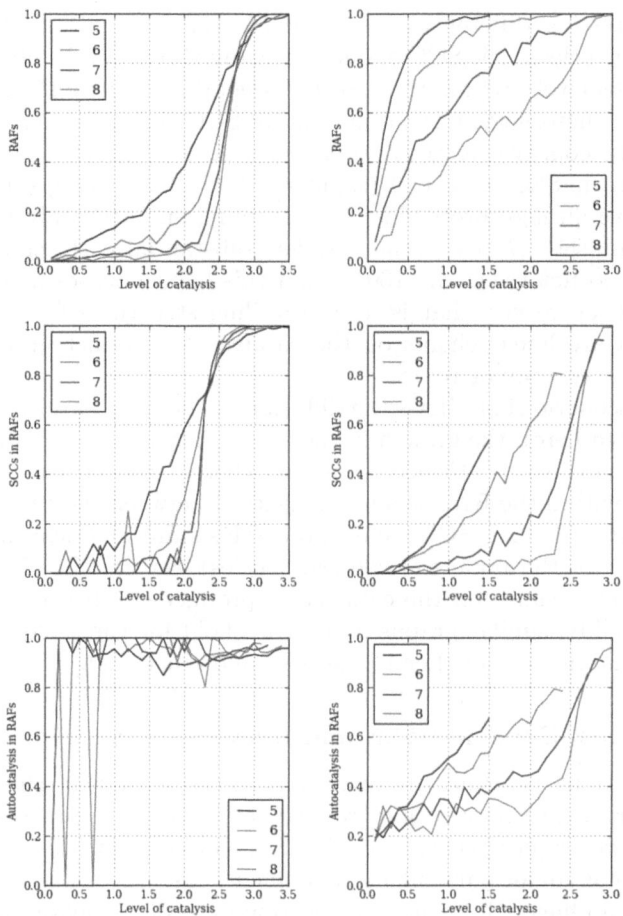

Fig. 4. TOP: Fraction of simulations showing at least 1 RAF. MIDDLE: fraction of RAFs having at least 1 SCC. BOTTOM: fraction of RAFs having at least 1 autocatalysis. On the x-axis $\langle c \rangle$ is represented, F is composed of all the species up to length 2 (left panel) and 3 (right panel). For each $\langle c \rangle$ and for each value of M, 1000 network instances have been created. For computational reasons, once the 100 % of networks with a specific $\langle c \rangle$ contain at least a RAF set, the system automatically goes to the next M, thus in some cases the analysis on SCC does not reach a $\langle c \rangle = 4$, the maximum level of catalysis evaluated.

final part of "MaxLen 8" scenarios: it is a clue that the line showing greatest chemistries (where there is a very low probability of choosing small molecules as catalysts) tends towards the plot of "BUF_2" scenario (where small molecules are not catalysts). Such ample systems (and the "BUF_2" scenario as well) may represent a good approximation of real chemistries.

More interesting differences are observed when comparing the fraction of systems having at least one SCC or one autocatalysis within the obtained RAFs.[8] At low catalysis levels, within the "BUF_2" scenario, almost all RAFs contain an autocatalysis, whereas the formation of other SCC structures inside the RAF is unlikely; on the contrary, as the average catalysis level grows up, the fraction of RAFs entailing an SSC tends to 1 (slightly before the 2.5 zone) - note that these RAFs still have an high probability of containing an autocatalysis as well. The situation within the "BUF_3" scenario differs substantially: the sum of SCCs and autocatalysis do not reach the 100 %, and this gap increases as the maximum allowed length decreases, that is, the prevailing structures for a large zone of catalysis level are linear chains on the catalyst → product graph. SCCs and autocatalysis play a minor role.[9]

Within a scenario that allows the diffusion of chemical species up to length 4 (data not shown here), the growth in the probability to observe a RAF happens even earlier than in the "BUF_3" scenario, whereas the presence of SCCs and autocatalysis within the RAF is almost identical between the two cases (indeed, the probability of having SCCs and autocatalysis are identical, not depending on the number of the buffered chemical species). So, in "BUF_4" scenario the presence of linear chains on the catalyst → product graph is even more significant. Remarkably, similar graphs and values hold for the scale-free topologies we tested (same number of chemical species and reactions).

3.3   Cleavages, Condensations and Irreversibility

Chemical reactions are generally reversible, in the sense that the reaction proceeds both forward (from reactants to products) and backward (in the opposite direction); however, note that, when far from equilibrium, the rates of the two processes are not equal. Particular environments (as aqueous phases rather than oleic phases) can hugely influence the reaction rates of all cleavages and/or of all condensations, whereas the presence or absence of catalysts can hugely influence the reaction rates of some specific reactions (for instance the particular sets of compartments, chemical species and reactions present inside the current living systems can hugely influence the reaction rates inside cells).

Hence, in case of similar rates the proportion between cleavages and condensations is 50 : 50, whereas in case of particular environments or particular living systems this proportion could constitute a relatively free parameter. In neutral environments the variation of this parameter do not have particular consequences: however, this is not the case where the concentration of small species is kept constant.

Indeed, since the membrane tends to allow the free transfer to small molecules only, the apparent symmetry of this situation is broken: whereas it is possible

---

[8] It is worth stressing that an autocatalysis is a strongly connected component, nevertheless we decided to deal with them separately.

[9] At high catalysis levels almost each RAF owns at least one SCC and one autocatalysis, so with our measurements it is not possible to observe the exact RAFs' structure. This aspect will be analyzed in further works.

to continuously construct long molecules by simply concatenating (by means of recursive condensation reactions) the always present small ones, it is not possible to continuously cleave long molecules, because their presence is not similarly guaranteed. Thus, it is not possible to observe RAF if just cleavages are present. Figure 5 shows the influence of different levels of the proportion between cleavages and condensations on the presence of RAF in chemical sets and on the position of their previously commented transition, in the case "BUF_2" scenario. In this case, note that, because the product of reactions involving buffered molecules are buffered molecules and because these buffered molecules do not catalyze anything, in presence of only cleavages nothing happen (data not shown here).

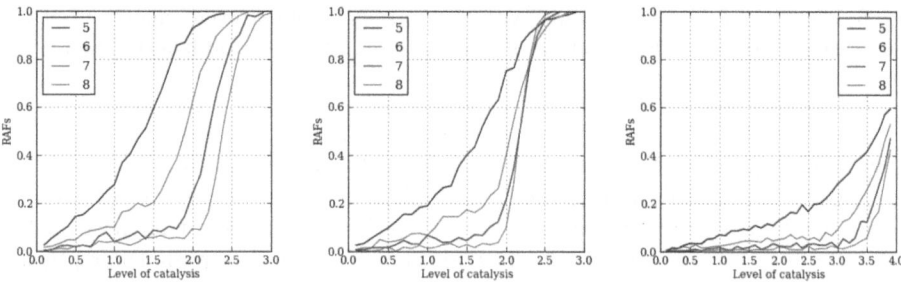

Fig. 5. The fraction of simulations with a food set composed of all the species up to length 2 showing at least 1 RAF, by varying the average level of catalysis and the maximum chemical species length. The plots show situations where the proportion between condensations and cleavages is 100 : 0 (left), all condensations; 75 : 25 (middle); 25 : 75 (right). The case 0 : 100 (all cleavages) is devoid of RAFs, hence graph is not shown here while the case 50 : 50 is shown in Fig. 4 left top.

Note that if we allow all the possible condensations and cleavages among the existing molecules up to length L, the relative proportion of cleavages and condensation is obtained by a purely combinatorial argument. n this case the number of conceivable reactions is equal to:

$$\hat{R}_1 = \sum_{i=2}^{L}[2^i \cdot (i-1)] + |S|^2 \tag{2}$$

where the first term stands for the cleavage reactions and the second term for the condensation reactions [12]. It is worth stressing that, as L increases and preventing the presence of reverse reactions, the number of conceivable condensation reactions tends to become higher than the number of conceivable cleavages, according to the ratio $[\sum_{i=2}^{L} 2^i \cdot (i-1)]/|S|^2$. In other words, the "purely combinatorial" proportion between cleavages and condensations is in favor of condensations, and this advantage increases with the diversity of the present chemical species.

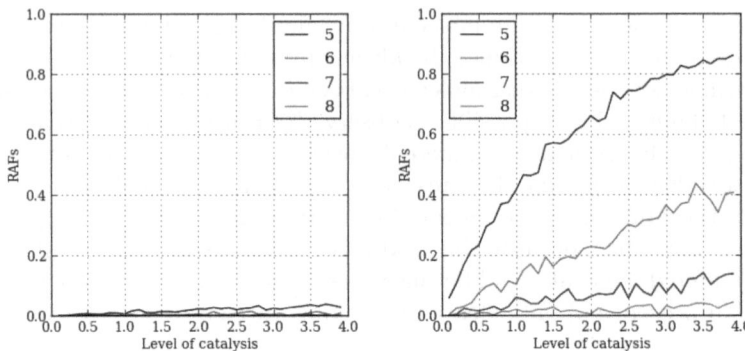

Fig. 6. Fraction of random systems, with the "purely combinatorial" quotient between cleavages and condensations, having at least 1 RAF, with food till length 2 and (c) till length 3

Following this idea, there may exist lots of condensations whose products are longer than L. This second method, applied to the same number of chemical species, leads to a very different number of reactions: so the question arises as how can we compare these two methods, and what can we say about their physical plausibility. We can observe that in (the relatively simple) pre-biotic environment very long molecules rarely appear, because of generic physical-chemical constraints. Thus, we think that the more correct model interpretation of the L limit should be: "chemical species longer than L are not suitable for further chemical processes, because of physical-chemical constraints" (for example, lack of solubility). A large fraction of the condensations leads therefore to not suitable products, which cannot play an active role within the system; thus the creation of this kind of "garbage" can inhibit the formation of RAFs (see Fig. 6).

4   Conclusion

The comprehension of the properties of sets of many interacting molecules poses formidable problems, and it is still a big challenge to obtain them in laboratory.

In this work we aim at revealing some properties of abstract realizations of catalytic reaction networks, with particular regard to the presence of structures potentially able to sustain their own existence and growth (the Reflexively Autocatalytic Food generated sets, briefly RAFs). Our investigations take into account different scenarios, involving different distributions of the chemicals' catalyzing capabilities, the presence or absence of backward reactions, and the presence or absence of long chemical species that lead to the accumulation of useless chemical species.

We confirmed that there is an ample region where the systems have autocatalytic structures, which are nevertheless unable to self-sustain (RAFs are not present); interestingly, this region shows the same amplitude across different topologies and/or settings. At a sufficient level of catalysis, complete RAFs emerge, whose presence and composition (different proportion of strongly connected components, autocatalysis and linear chains) depend on some parameters (average level of catalysis, number of buffered chemical species, proportion between cleavages and condensations) and is independent on other characteristics (uniform or scale-free topology, presence or absence of backward reactions). Finally, we discussed two different ways to build artificial networks and their physical implications.

We think that these hints could be useful for the comprehension of the emergence of sets of molecules able to collectively self-replicate in OoL scenarios, and for the designing of new artificial protocells.

Acknowledgements. The research leading to these results has received funding from the European Union Seventh Framework Programme (FP7/2007–2013) under grant agreement n. 284625 and from "INSITE - The Innovation Society, Sustainability, and ICT" Pr.ref. 271574, under the 7th FWP - FET programme.

References

1. Alberts, B., Johnson, A., Lewis, J., Raff, M., Roberts, K., Walter, P.: Molecular Biology of the Cell, 4, 5 edn. Garland Science, New York (2005)
2. Bagley, R., Farmer, J.D.: Spontaneous emergence of a metabolism. In: Langton, C. (ed.) Artificial Life II. Santa Fe Institute Studies in the Sciences of Complexity, vol. X, pp. 93–141. Addison-Wesley, Redwood (1992)
3. Barabási, A.-L., Albert, R.: Emergence of scaling in random networks. Science **286**(5439), 509–512 (1999)
4. Carletti, T., Serra, R., Villani, M., Poli, I., Filisetti, A.: Sufficient conditions for emergent synchronization in protocell models. J. Theor. Biol. **254**(4), 741–751 (2008)
5. Diester, R.: Graph Theory. Springer, Heidelberg (2005)
6. Dyson, F.: Origins of Life. Cambridge University Press, Cambridge (1985)
7. Eigen, M., Schuster, P.: The hypercycle. a principle of natural self-organization. part A: Emergence of the hypercycle. Die Naturwiss. **11**(64), 541–565 (1977)
8. Eigen, M., Schuster, P.: The Hypercycle: A Principle of Natural Self-Organization. Springer, New York (1979)
9. Farmer, J., Kauffman, S.: Autocatalytic replication of polymers. Phys. D **220**, 50–67 (1986)
10. Filisetti, A., Graudenzi, A., Serra, R., Villani, M., De Lucrezia, D., Fuchslin, R.M., Kauffman, S.A., Packard, N., Poli, I.: A stochastic model of the emergence of autocatalytic cycles. J. Syst. Chem. **2**(1), 2 (2011)
11. Filisetti, A., Graudenzi, A., Serra, R., Villani, M., De Lucrezia, D., Poli, I.: The role of energy in a stochastic model of the emergence of autocatalytic sets. In: Lenaerts, T., Giacobini, M., Bersini, H., Bourgine, P., Dorigo, M., Doursat, R. (eds.) Proceedings of the Eleventh European Conference on the Synthesis and Simulation of Living Systems Advances in Artificial Life, ECAL 2011, pp. 227–234. MIT Press, Cambridge (2011)
12. Filisetti, A., Graudenzi, A., Serra, R., Villani, M., Füchslin, R.M., Packard, N., Kauffman, S.A., Poli, I.: A stochastic model of autocatalytic reaction networks. Theory in biosciences = Theorie in den Biowissenschaften, pp. 1–9, Oct 2011

13. Filisetti, A., Serra, R., Carletti, T., Villani, M., Poli, I.: Non-linear protocell models: synchronization and chaos. Eur. Phys. J. B **77**(2), 249–256 (2010)
14. Filisetti, A., Serra, R., Villani, M., Graudenzi, A., Füchslin, R.M., Poli, I.: The influence of the residence time on the dynamics of catalytic reaction networks. In: Frontiers in Artificial Intelligence and Applications - Neural Nets WIRN10 - Proceedings of the 20th Italian Workshop on Neural Nets, pp. 243–251 (2011)
15. Ganti, T.: Chemoton Theory, vol. I: Theory of fluyd machineries: vol. II: Theory of livin system. Kluwer Academic/Plenum, New York (2003)
16. Hordijk, W., Kauffman, S.A., Steel, M.: Required levels of catalysis for emergence of autocatalytic sets in models of chemical reaction systems. Int. J. Mo. Sci. **12**(5), 3085–3101 (2011)
17. Hordijk, W., Steel, M.: Detecting autocatalytic, self-sustaining sets in chemical reaction systems. J. Theor. Biol. **227**(4), 451–461 (2004)
18. Hordijk, W., Steel, M., Kauffman, S.: The Structure of Autocatalytic Sets: Evolvability, Enablement, and Emergence, p. 12, May 2012
19. Jain, S., Krishna, S.: Autocatalytic set and the growth of complexity in an evolutionary model. Phys. Rev. Lett. **81**, 5684–5687 (1998)
20. Jain, S., Krishna, S.: A model for the emergence of cooperation, interdependence, and structure in evolving networks. Proc. Natl. Acad. Sci., USA **98**(2), 543–547 (2001)
21. Kauffman, S.A.: Autocatalytic sets of proteins. J. Theor. Biol. **119**(1), 1–24 (1986)
22. Kauffman, S.A.: The Origins of Order. Oxford University Press, New York (1993)
23. Luisi, P., Ferri, F., Stano, P.: Approaches to semi-synthetic minimal cells: a review. Naturwissenschaften **93**(1), 1–13 (2006)
24. Mansy, S.S., Schrum, J.P., Krishnamurthy, M., Tobé, S., Treco, D.A., Szostak, J.W.: Template-directed synthesis of a genetic polymer in a model protocell. Nature **454**(7200), 122–125 (2008)
25. Mossel, E., Steel, M.: Random biochemical networks: the probability of self-sustaining autocatalysis. J. Theor. Biol. **233**(3), 327–336 (2005)
26. Rasmussen, S., Chen, L., Nilsson, M., Abe, S.: Bridging nonliving and living matter. Artif. Life **9**(3), 269–316 (2003)
27. Rocheleau, T., Rasmussen, S., Nielsen, P.E., Jacobi, M.N., Ziock, H.: Emergence of protocellular growth laws. Philos. Trans. R. Soc. Lond. B. Biol. Sci. **362**(1486), 1841–1845 (2007)
28. Schmidt, L.: The Engineering of Chemical Reactions. Oxford University Press, New York (1998)
29. Segre, D., Lancet, D., Kedem, O., Pilpel, Y.: Graded autocatalysis replication domain (gard): kinetic analysis of self-replication in mutually catalytic sets. Orig. Life Evol. Biosph. **28**, 501–514 (1998)
30. Serra, R., Carletti, T., Poli, I.: Syncronization phenomena in surface-reaction models of protocells. Artif. Life **123**(2), 123–138 (2007)
31. Serra, R., Filisetti, A., Villani, M., Graudenzi, A., Damiani, C., Panini, T.: A stochastic model of catalytic reaction networks in protocells. Nat. Comput. (2014). doi:10.1007/s11047-014-9445-6
32. Solé, R.V., Munteanu, A., Rodriguez-Caso, C., Macía, J.: Synthetic protocell biology: from reproduction to computation. Philos. Trans. R. Soc. Lond. B. Biol. Sci. **362**(1486), 1727–1739 (2007)
33. Stadler, P.F.: Dynamics of autocatalytic reaction networks. IV: Inhomogeneous replicator networks. Bio. Syst. **26**(1), 1–19 (1991)
34. Szostak, J.W., Bartel, D.P., Luisi, P.L.: Synthesizing life. Nature **409**(6818), 387–390 (2001)

Learning Multiple Conflicting Tasks
with Artificial Evolution

Delphine Nicolay[1]([⊠]), Andrea Roli[2], and Timoteo Carletti[1]

[1] Department of Mathematics and NaXys, University of Namur, Namur, Belgium
delphine.nicolay@unamur.be
[2] Department of Computer Science and Engineering (DISI), University of Bologna,
Campus of Cesena, Bologna, Italy

Abstract. The paper explores the issue of learning multiple competing tasks in the domain of artificial evolution. In particular, a robot is trained so as to be able to perform two different tasks at the same time, namely a gradient following and rough terrain avoidance behaviours. It is shown that, if the controller is trained to learn two tasks of different difficulty, then the robot performance is higher if the most difficult task is learnt first, before the combined learning of both tasks. An explanation to this superiority is also discussed, in comparison with previous results.

1 Introduction

Since its first appearance in the late 60's, artificial evolution (AE) has been successfully applied to solve a plethora of problems arising in fields such as circuit design, control systems, data mining, optimisation and robotics. The application of AE to robotics is commonly named *evolutionary robotics* (ER) and it is mainly concerned with the artificial evolution of robots neural controllers [12]. ER has contributed with notable results to the domain of artificial cognition and it also provides a viable way for designing robotic systems, such as robot swarms. Nevertheless, in spite of promising initial results, ER has not completely expressed its potential, especially in the design of complex robotic controllers. In fact, ER is still applied with large contributions of the experience and the intuition of the designer. To close the gap between current ER achievements and its potential applications, both theoretical and methodological advancements are required.

A case in point is the issue of training a robot controller such that the robot is able to accomplish more than one task at the same time. The practice in robotic learning already possesses some good practice and specific methodologies, such as robot shaping [5], which is primarily applied to incremental training of robots. However, general results in this context are still lacking.

A prominent study is that by Calabretta et al. [3], in which the question as to whether a non-modular neural network can be trained so as to achieve the same performance as a modular one in accomplishing multiple tasks is addressed. The neural network is trained to perform two simultaneous classification tasks on the same input, one task being harder than the other. The conclusion of the

© Springer International Publishing Switzerland 2014
C. Pizzuti and G. Spezzano (Eds.): WIVACE 2014, CCIS 445, pp. 127–139, 2014.
DOI: 10.1007/978-3-319-12745-3_11

results of these experiments is particularly interesting: the best performance is attained by training first on the harder tasks and successively also on the second. This learning sequence is the one enabling the search process to minimise the so-called *neural interference*, which appears every time neural weights involved in the computation of both tasks are varied considerably, leading to a detriment of the overall network performance. If the harder task is learnt first, then neural interference is reduced as much as possible, because the magnitude of weight variation is kept at minimal levels. The work by Calabretta and co-authors is particularly meaningful because it sheds light to the mechanisms underpinning the learning sequence that leads to the best performance—which is comparable to the one attained by a modular neural network, trained independently on the two tasks.

In this work, we address the question as to which is the best task learning sequence for training a robot. The case under study is more general than that in [3], as it consists of two conflicting robotic tasks—which are computationally more complex than classification tasks—and both network structure and weights can undergo changes during the training phase, which is based on a genetic algorithm (GA). We will show that the best results can be attained by training the robot on the harder task first and then on both tasks, confirming the results in [3]. Nevertheless, since our controllers and the training process are different than those in the study by Calabretta et al., the explanation based on neural interference is no longer valid and a different mechanism should be found. We conjecture that in the case of GA, the interference between the two tasks induces a rugged fitness landscape: learning first on the harder task means to keep the landscape as smooth as possible, thus enabling the search mechanism of GA to attain a higher performance than in the other learning cases.

In Sect. 2, we illustrate the scenario and two tasks the robot has to perform. In Sect. 3, we detail the experiments we made and we describe the results of the three possible training settings for the robot. Section 4 presents a discussion on the results and Sect. 5 concludes the contribution with an outline of future work.

2   Model and Tasks Description

The abstract application that we focus on follows ideas presented by Beaumont [1]. Virtual robots, controlled by neural networks, are trained to learn two different tasks in a virtual arena. This arena is a discretised grid of varying sizes, hereby for a sake of simplicity we fixed it to 20×20, with a torus topology, i.e. periodic boundary conditions on both directions, on which robots are allowed to move into any neighbouring square at each displacement, that is the 8–cells Moore neighbourhood. Assuming the arena is not flat and it possesses one global maximum, the first task we proposed involves reaching this global maximum, represented using a colour code in the left panel of Fig. 1; the second task consists in moving as long as possible in the arena by avoiding "dangerous" zones, i.e. zones where robots loose energy, represented by black boxes in the right panel of Fig. 1.

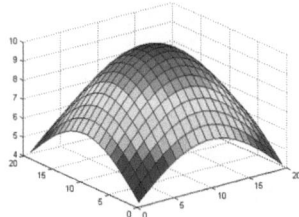 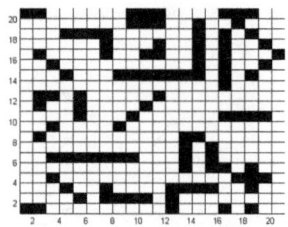

Fig. 1. Arenas where robots are trained. Left panel: arena for task T_1, the peak level is represented as a color code: red high level, blue lower levels. Right panel: arena for task T_2, black boxes denote the "dangerous" zones to be avoided as much as possible.

The model of neural networks [13,14] that we use is schematically shown in Fig. 2. The topology is completely unconstrained, except for the self–loops that are prohibited. Each neuron can have two internal states: activated (equal to 1) or inhibited (equal to 0). The weight of each link is 1, if the link is excitatory, or −1 if it is of inhibitory nature. Given the present state for each neuron, (x_i^t), the updated state of each neuron is given by the perceptron rule:

$$\forall j \in \{1, \ldots, N\} \quad : \quad x_j^{t+1} = \begin{cases} 1 & \text{if } \sum_{i=1}^{k_j^{in}} w_{ji}^t x_i^t \geq \theta k_j^{in} \\ 0 & \text{otherwise} \end{cases}$$

where k_j^{in} is the number of incoming links in the j–th neuron under scrutiny, θ the threshold of the neuron, a parameter lying in the interval $[-1,1]$, and w_{ji}^t is the weight, at time t, of the synapse linking i to j.

The robots that we consider are built using 17 sensors and 4 motors. The 9 first sensors help robots to check the local height of the square on which the robot is with respect to the squares around it. The 8 other ones are used to detect the presence of "dangerous" zones around it. The 4 motors control the basic movements of robots, i.e. north–south–east and west, and combining them, they also allow diagonal movements. Hence, the neural networks that control the robots are made of 43 nodes: 17 inputs, 4 outputs and 22 hidden neurons. This number has been set large enough to let sufficient possibilities of connections in order to achieve the tasks, but still keeping reasonably low not to use too much CPU resources.

As the state of each neuron of our neural networks is binary, we transform the information acquired by the sensors in binary inputs to introduce them in the networks. Consequently, the state x_i of the nine first inputs is calculated according to the formula $x_i = \left\lfloor \frac{1.9(h_i - h_i^{min})}{h_i^{max} - h_i^{min}} \right\rfloor$ where h_i is the height of sensor i and h_i^{min} and h_i^{max} respectively represent the minimal and maximal heights among the 9 available ones for the robot. For the last eight inputs responsible for the exploration of the neighbouring zones, their state is 1 if their associated sensor detects a "dangerous" square and 0 otherwise.

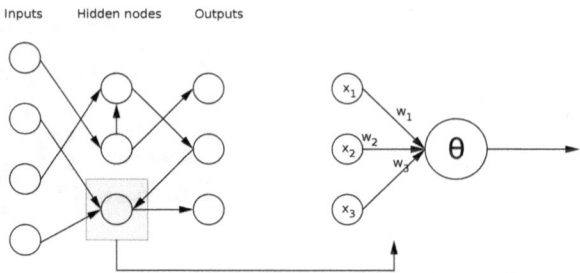

Fig. 2. Schematic presentation of the artificial neural networks model used in our research. The topology is unconstrained and the state of each neuron is a binary value. The updated state of each neuron is given by the perceptron rule.

The binary neural network outputs are then transformed back into motors' movements, more precisely we devote two motors to the west-east movements and the two others to the north-south movements. The robot movements follow the rule detailed in Table 1.

Table 1. Movement achieved by the robots according to the network's outputs. The 8–cells Moore neighbourhood can be reached by the robots at each displacement.

Output 1	Output 2	Movement
0	0	-1
0	1	0
1	0	0
1	1	$+1$

For each task, the quality of robot's behaviour is quantified in order to determine the network that performs the best control. In the estimation of the first task, i.e. reaching the global maximum of the arena, the robot carries out 20 steps for each of the 10 initial positions of the training set. For each of these displacements, we save the mean height of the last five steps. Then, we calculate a fitness value, fit, which is the mean height on all the displacements. The quality of the robot behaviour is finally given by $quality_1 = \frac{fit - h_{min}}{h_{max} - h_{min}}$ where $h_{min} = 3.93$ and $h_{max} = 9.98$ are respectively the minimal and maximal heights of the grid. A robot, and so a network, has a quality of 1 if, for each initial position, it reaches the peak at least at the sixteenth step and stays on this peak until the end of the displacement.

For the assessment of the second task, namely avoiding "dangerous" zones, the robot is assigned[1] an initial energy equal to 100 for the 5 initial positions in

---

[1] Let us observe that such parameters have been set to the present values after a careful analysis of some generic simulations and by a test and error procedure.

each of the 3 configurations of the training set. For these fifteen cases, it performs 100 steps in the arena. If it moves to a "dangerous" square, its energy is decreased by 5. Moreover, if it moves to a square that it has already visited, its energy is reduced by 1. This condition is needed to force the robot to move instead of remaining on a safe cell. We save the final energy of each case and we compute the mean final energy. If the energy becomes negative, we set its value to zero. The quality of the robot's behaviour is then calculated as $quality_2 = \frac{\text{final energy}}{\text{maximum energy}}$. A robot has a quality of 1 if, for each initial position of each configuration, it moves continuously on the grid without visiting "dangerous" squares and squares already visited.

When both tasks are performed simultaneously, we slightly adjust our training and our quality measures. The training set is composed by the three configurations of the second task and their five respective initial positions. The robot is assigned an energy equal to 20 instead of 100 and it performs 20 steps in the arena. The rule for the loss of energy remains the same as well as the way of calculating the mean height on all the displacements. Nevertheless, when the final energy of one displacement is zero, we assigned a value of zero to its mean height on the last five steps. If the value fit is lower than h_{min}, we set the quality measure to zero. This modification in the quality of the fitness is introduced to prevent robots from having a very good behaviour for one task and a very bad behaviour for the other one. Let us note that these two tasks are competing each other. Indeed, in the first task, we hope that robot stops on the peak and in the second one, we ask it to move continuously. So, we look for a trade-off between the two tasks in the robot behaviour.

The suitable topology and the appropriate weights and thresholds have to be fine tuned to obtain neural networks responsible for good performance and this strongly depends on the given arena, however we are interested in obtaining a robust behaviour irrespectively from the chosen environment, that is why we resort to heuristic optimisation method, namely genetic algorithms (GA), that have been used abundantly in the literature [4,6,10]. The choice of making use of genetic algorithms as optimisation method has been made for two reasons. Firstly, the perceptron rule is a discontinuous threshold function and we can not directly apply some well-known algorithms such as backpropagation. Secondly, GA optimises both topology and weights–thresholds, which is necessary as the topology of our networks is not frozen. Some preliminary results about these heuristic methods have already been published in [11].

The genetic algorithm used in the following is discrete-valued for the weights and real-valued for the thresholds. The selection is performed by a roulette wheel selection. The operators chosen are the classical 1-point crossover and 1-inversion mutation. Their respective rates are 0.9 and 0.01, while the population size is 100 and the maximum number of generations is 10 000. To ensure the legacy of best individuals, the population of parents and offsprings are compared at each generation before keeping the best individuals among both populations. New random individuals (one-tenth of the population size) are also introduced at each generation by replacing worst individuals in order to avoid premature convergence.

3   Results

In a first stage, we performed each task separately to test their feasibility. Then, we compared networks trained on each task learnt successively and networks trained on both task simultaneously; as well as different learning sequences in order to discover the best way to train robots for multiple tasks.

For each learning process, we performed 10 trainings in order to get valuable results, given the heuristic characteristic of our optimisation method. We used computation resources of an intensive computing platform. Indeed, one optimisation is time consuming. The cost of our algorithm is due to the evaluation of the quality performance. For example, the time for one evaluation of the first and the second task quality measure are respectively 0.0192 and 0.0626 seconds for a processor Intel Core i5–3470 CPU @ 3.20 GHz x 4 and a memory of 7,5 Gio. About 1100000 evaluations are performed during an optimisation and our algorithm is sequential, which means that one training takes approximatively 5 hours for the first task and 19 hours for the second one. All our codes are written in Matlab.

3.1   Single Tasks

Figure 3 represents the maximum performance attained at each generation of one optimisation for reaching the global maximum of the arena on the left panel and for avoiding "dangerous" zones on the right panel. We can remark that the second task is the most difficult to achieve. Indeed, we observe that the optimisation of reaching the global maximum is perfectly performed in less than 250 generations. However, avoiding "dangerous" zones does not reach the maximum performance at the end of the optimisation (after 10 000 generations).

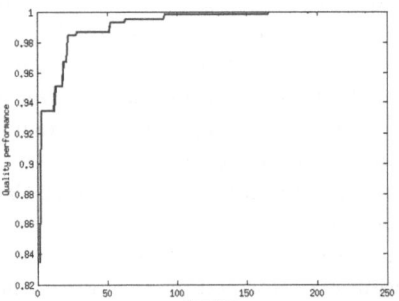 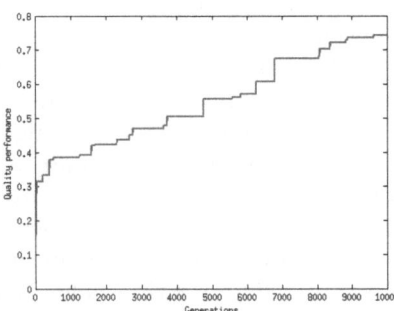

Fig. 3. Maximum performance reached at each generation of the genetic algorithm for the achievement of one particular task. Left panel: maximum performance for reaching the global maximum of the arena. Right panel: maximum performance for avoiding "dangerous zones". The achievement of the first task is perfect after a few generations while the achievement of the second task does not reach the maximal performance.

3.2 Combined Tasks

Three different learning processes are considered in order to find the best way to train robots for multiple tasks. In the first case, networks learn both tasks simultaneously giving to the total quality measure an equal weight (named in the following 0.5T1 + 0.5T2). In the second one, robots are first trained to achieve the first task and then both tasks simultaneously (named, T1 → T1+T2) while for the last sequence, it is the second task that is trained before the simultaneous training (T2 → T1+T2). During the simultaneous learning, the quality performance is once again the average of the quality measures of T1 and T2.

Remark 1. Let us observe that a fourth possible configuration could be used, that is, to consider two smaller networks having inputs only for a given task[2] and then juxtapose the two and link in couple each motor output for each small network, in such a way to get a larger network. Some preliminary results showed that the network obtained by juxtaposition of two smaller ones, each one optimised for a task, does not seem to have good performance. In the following we will not consider this fourth case because in our opinion these results need to be further analysed, before any conclusion could be done.

Table 2 presents, for each learning sequence, the mean and standard deviation of the final performance computed from ten independent realisations. The non-parametric hypothesis test of Kruskal-Wallis [7] is considered to evaluate the difference of medians. As the p-value is equal to 0.1232, we accept the null hypothesis, which means that the medians of the three learning sequences seem to be similar on the training set.

Table 2. Mean and standard deviation, on 10 optimisations, of the final maximum performance got for each learning sequence. The total quality measure is calculated as the average of the quality performance of T1 and T2 for all the three learning processes.

Learning process	Mean	Std
0.5T1 + 0.5T2	0.7295	0.0588
T1 → T1+T2	0.7252	0.0407
T2 → T1+T2	0.7606	0.0606

In order to properly assess the quality of the controllers obtained at the end of the training process, we compare their behaviour using validation sets, i.e., scenarios not present in the set used for the training. We consider different ways of changing the arena. In the first and second validation set, we modify respectively the zones where robots lose energy and the shape of the surface to

---

[2] More precisely, one network composed by 9 inputs, 5 hidden neurons and 4 outputs, has only the inputs able to measure the local heights and the second one, formed by 8 inputs, 5 hidden neurons and 4 outputs, to detect the dangerous zones.

Table 3. Mean and standard deviation of the performance obtained by random neural networks and by the neural networks trained with the different learning sequences on three validation sets, each of them representing a different way to modify the arena.

Area	Measure	Random network	0.5T1 + 0.5T2	T1 → T1+T2	T2 → T1+T2
Validation	Mean	0.0047	0.1073	0.1520	0.2718
1	Std	0.0036	0.0707	0.1073	0.1290
Validation	Mean	0.0120	0.2946	0.3379	0.4318
2	Std	0.0060	0.1259	0.1037	0.1593
Validation	Mean	0.0053	0.1746	0.1687	0.3313
3	Std	0.0076	0.1233	0.0933	0.1390

climb (position of the peak and slope of the surface). In the last set, we perform both modifications. Table 3 presents, for each of these validation sets, the mean and standard deviation of the performance got by the neural networks trained with the different learning sequences. A boxplot of the results obtained for each learning sequence is presented on the right panel of Fig. 4.

The third learning sequence, i.e. the training of the second task followed by the one of both tasks (T2 → T1+T2), seems to be the best way to train robots for multiple tasks. Indeed, the mean performance of this sequence is the highest for each validation set. Moreover, this learning sequence increases the mean performance by at least 25 % compared to other learning sequences. But we also note that the standard deviation is the highest for this learning process.

To confirm our intuition, we make use of statistical tests whose results are presented on the left panel of Fig. 4. the first test that we use is Kruskal-Wallis in order to check the equality of the three medians. As the p-value is smaller than 0.05, we come to the conclusion that the medians are significantly different. Then, we use the non-parametric Wilcoxon test [7] to evaluate medians by pair.

Kruskal-Wallis

p-value	
$3.8281.10^{-4}$	< 0.05

Wilcoxon

Seq. 1	Seq. 2	p-value	
0.5T1 + 0.5T2	T1 → T1+T2	0.4161	> 0.05
0.5T1 + 0.5T2	T2 → T1+T2	0.0002	< 0.05
T1 → T1+T2	T2 → T1+T2	0.0033	< 0.05

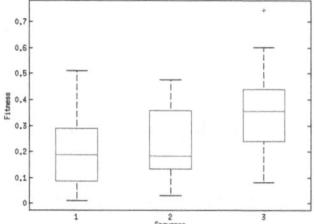

Fig. 4. Analysis of the quality values obtained for the three learning sequences. The performance is evaluated on the validation sets by the optimal networks of 10 optimisations. Left panel: hypothesis tests of Kruskal-Wallis and Wilcoxon used to compare the medians of the quality performance. Right panel: boxplot of the quality measures.

Table 4. Mean and standard deviation of the task 1 performance obtained by the neural networks trained with the different learning sequences on three validation sets, each of them representing a different way to modify the arena.

Area	Measure	0.5T1 + 0.5T2	T1 → T1+T2	T2 → T1+T2
Validation	Mean	0.0520	0.1296	0.2672
1	Std	0.0879	0.1441	0.1995
Validation	Mean	0.3322	0.4305	0.4625
2	Std	0.1667	0.1736	0.1872
Validation	Mean	0.2063	0.1910	0.3937
3	Std	0.2085	0.1401	0.2240

Table 5. Mean and standard deviation of the task 2 performance obtained by the neural networks trained with the different learning sequences on three validation sets, each of them representing a different way to modify the arena.

Area	Measure	0.5T1 + 0.5T2	T1 → T1+T2	T2 → T1+T2
Validation	Mean	0.1627	0.1743	0.2763
1	Std	0.0770	0.0771	0.0748
Validation	Mean	0.2570	0.2453	0.4010
2	Std	0.1102	0.0594	0.1554
Validation	Mean	0.1430	0.1463	0.2630
3	Std	0.0589	0.0590	0.0646

With this test, we observe that sequence 1 and 2 leads to similar medians while the one of sequence 3 is significantly different. Let us remind that the second task, which is first trained in this sequence, is the most difficult to learn. In conclusion, it seems better to train the hardest task before training both tasks simultaneously.

Tables 4 and 5 present the performance associated with the first and the second task, respectively. We can see that the learning sequence T2 → T1+T2 makes it possible to attain the best results because it helps the training process to improve the fitness function of both the tasks.

4   Discussion

The result of our study about learning sequences is that it seems to be better to learn the hardest task before learning both tasks. This conclusion coincides with the one drawn by Calabretta et al. Nevertheless, far from appearing as a replicate of this study, we flesh out the spectrum of experiments on which the results are valid. Indeed, it is worthwhile to summarise the main differences between our context of study and the one in [3]. Firstly, the two tasks considered in this study are dynamical tasks rather than classification ones. Secondly, both weights and structure of the networks undergo changes during learning. Thirdly, the tasks have separate inputs but common outputs, which is the opposite of Calabretta's experiments. Finally, let us observe that the considered tasks are conflicting and that the weights of the networks do not change in magnitude.

Although our study leads to the same conclusion of the one in [3], we can not use the same reason to explain it. Indeed, since the networks and the learning process are different, the neural interference is no longer valid and a different mechanism should be found. To this aim, we study the autocorrelation of our performance landscape [8] for each learning sequence. For this, we perform 10 random walks of 10 000 steps in all learning schemes presented before. These walks are executed by random networks for the separate and simultaneous learning of tasks (T1, T2 and 0.5T1 + 0.5T2) while they are performed by networks trained for task 1 and task 2 respectively in the case of successive learning (T1 → T1+T2 and T2 → T1+T2). Figure 5 presents the mean of the autocorrelation on the 10 walks for each length from 1 to 5 and for each learning sequence on the left panel and the mean and standard deviation for autocorrelation of length 1 in the table on the right.

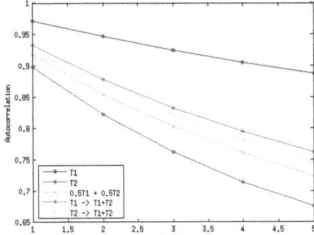

Learning sequence	Mean	Std
T1	0.9717	0.0061
T2	0.8981	0.0192
0.5T1 + 0.5T2	0.9187	0.0122
T1 → T1+T2	0.9331	0.0181
T2 → T1+T2	0.9253	0.0184

Fig. 5. Study of the autocorrelation of the performance landscape. Left panel: mean of the autocorrelation on 10 walks for each length from 1 to 5 and for each learning sequence. Right panel: Mean and standard deviation of the autocorrelation of length 1 for each learning process.

The autocorrelation values explain the marked differences in the final performance of the robot in the extreme cases, i.e., task 1—the easiest, with the highest autocorrelation values corresponding to a flatter fitness landscape—and task 2—the hardest, with the lowest values and thus a more rough fitness landscape. Moreover, the learning sequence 0.5T1 + 0.5T2 clearly corresponds to autocorrelation values just above the ones of T2, but below the others. This explains why this sequence is the worst among the three ones with the compound task. The final high performance attained by the learning sequence T2 → T1+T2 does not seem to find support in the autocorrelation values, which are quite close. However, we must observe that the measure is taken on a landscape corresponding to the final phase of evolution on task 1 (T1 → T1+T2) and task 2 (T2 → T1+T2), respectively. Therefore, at the beginning of the search with the composite fitness (accounting for both the tasks), the autocorrelation of the landscape is strongly influenced by the initial training phase with only one task active. Hence, the landscape autocorrelation in the sequence T2 → T1+T2 heavily depends on the autocorrelation of the landscape for learning task 2, which is the very low. We can conclude that in the T2 → T1+T2, the addition of the fitness component corresponding to task 1 is considerably beneficial to the search process, enabling the genetic algorithm to explore a way smoother landscape, whilst the addition of the task 2 fitness has a dramatically detrimental effect, harming search effectiveness.

The second point that we discuss is the existence of correlation between the tasks. With this aim, we perform the removal of one hidden neuron at once and we analyse the modifications of the quality measures. These modifications are shown in Fig. 6 for 10 optimisations of the networks on both tasks simultaneously. We can observe a linear relation between the modifications, which is confirmed by the correlation coefficient (0.8138). However, can we conclude that neurons that are important for the first task are also important for the second one? Or, is this correlation due to the important number of links in the networks, that we modify too strongly by removing one neuron?

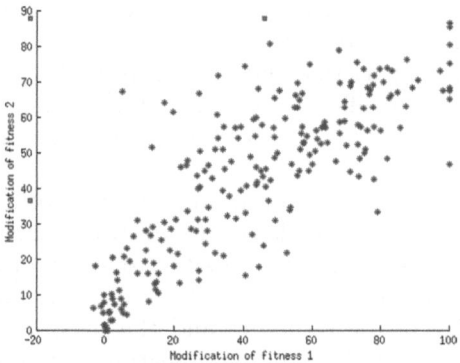

Fig. 6. Relation between the modifications of the two tasks when a hidden neuron is removed from the network. Networks that are considered are the ones obtained after the training on both tasks simultaneously (0.5T1 + 0.5T2).

To provide a first answer to this question we run the same kind of experiments but with a definitive removal of respectively 3 or 5 random chosen hidden nodes before temporarily removing each of the remaining ones and we observed that the correlation between the tasks decreases.

In order to learn more about the topology of our networks and the relations between the two tasks, we need to study the modularity of our networks. Two kinds of modularities can be considered. The first one is a topological modularity [9] that can be studied with some community detection methods such as the Louvain method [2] or by using measures of centrality or similarity. In order to analyse the second one, namely the functional modularity, that is to group together neurons that have similar dynamic behaviours, we plan to use some recent information-like indicators [15]. This study will be done in future work.

5   Conclusion

Evolutionary robotics has led to important progress in the domain of artificial cognition. Nevertheless, general results are still lacking about the issue of training a robot controller such that the robot accomplishes multiple tasks. This paper has examined what is the best learning sequence for the training of two tasks with different difficulty. The conclusion of this study has expanded the results brought to light by Calabretta et al. [3] to a more general framework.

In this work, we train robot controllers such that the robot is able to reach the maximum of a surface while avoiding "dangerous" zones where it looses energy. The optimisation is executed both on weights and structure of the networks. Different learning sequences are envisaged in order to find the best way for learning multiple tasks.

The learning sequence that has been shown to be appropriate for the learning of conflicting multiple tasks is the sequence in which the most difficult task is learnt before the learning of both tasks simultaneously. Contrary to the study of Calabretta, we can not ascribe the difference in the performance to neural interference, because of a different experimental setting. However, a possible explanation can come from the study of the autocorrelation of the fitness landscape.

The conclusion of this study is promising and supports the claim about the generality of the result, as it extends the application framework with respect to previous works. Further work will address the topology of the networks as well as the relations between the two tasks by the study of the networks modularity. This better knowledge of the structure and the tasks relations could lead to important progress in the understanding of learning process.

Acknowledgements. This research used computational resources of the "Plateforme Technologique de Calcul Intensif (PTCI)" located at the University of Namur, Belgium, which is supported by the F.R.S.-FNRS.

This paper presents research results of the Belgian Network DYSCO (Dynamical Systems, Control, and Optimisation), funded by the Interuniversity Attraction Poles Programme, initiated by the Belgian State, Science Policy Office. The scientific responsibility rests with its authors.

References

1. Beaumont, M.A.: Evolution of optimal behaviour in networks of boolean automata. J. Theor. Biol. **165**, 455–476 (1993)
2. Blondel, V.D., Guillaume, J.-L., Lambiotte, R., Lefebvre, E.: Fast unfolding of communities in large networks. J. Stat. Mech: Theory Exp. **2008**(10), P10008 (2008)

3. Calabretta, R., Di Ferdinando, A., Parisi, D., Keil, F.C.: How to learn multiple tasks. Biol. Theor. **3**(1), 30–41 (2008)
4. Deb, K.: Multi-Objective Optimization using Evolutionary Algorithms. John Wiley & Sons Ltd, Chichester (2008)
5. Dorigo, M., Colombetti, M.: Robot Shaping: An Experiment in Behavior Engineering. The MIT Press, Cambridge (1998)
6. Goldberg, D.E.: Genetic Algorithms in Search, Optimization, and Machine Learning. Addison-Wesley, Boston (1989)
7. Hollander, M., Wolfe, D.A., Chicken, E.: Nonparametric statistical methods, vol. 751. John`Wiley & Sons, New York (2013)
8. Hoos, H.H., Stützle, T.: Stochastic local search: Foundations & applications. Elsevier (2004)
9. Kashtan, N., Alon, U.: Spontaneous evolution of modularity and network motifs. Proc. Nat. Acad. Sci. U.S.A. **102**(39), 13773–13778 (2005)
10. MacKay, D.: Information Theory, Inference, and Learning Algorithms. Cambridge University Press, Cambridge (2005)
11. Nicolay, D., Carletti, T.: Neural networks learning: some preliminary results on heuristic methods and applications. In: Perotti, A., Di Caro, L. (eds.) DWAI@AI*IA, volume 1126 of CEUR Workshop Proceedings, pp. 30–40 (2013). CEUR-WS.org
12. Nolfi, S., Floreano, D.: Evolutionary robotics. The MIT Press, Cambridge (2000)
13. Peretto, P.: An introduction to the modeling of neural networks, vol. Alea Saclay. Cambridge University Press, Cambridge (1992)
14. Rojas, R.: Neural networks: A systematic introduction. Springer, Berlin (1996)
15. Villani, M., Filisetti, A., Benedettini, S., Roli, A., Lane, D.A., Serra, R.: The detection of intermediate-level emergent structures and patterns. In: Advances in Artificial Life, ECAL, vol. 12, pp. 372–378 (2013)

On Some Properties of Information Theoretical Measures for the Study of Complex Systems

Alessandro Filisetti[1]([✉]), Marco Villani[1,2], Andrea Roli[1,3], Marco Fiorucci[1,4], Irene Poli[1,4], and Roberto Serra[1,2]

[1] European Centre for Living Technology, Ca' Foscari University of Venice,
Venice, Italy
alessandro.filisetti@unive.it
[2] Department of Physics, Informatics and Mathematics, University of Modena
and Reggio Emilia, Modena, Italy
[3] Dept. of Computer Science and Engineering (DISI),
University of Bologna, Bologna, Italy
[4] Department of Environmental Sciences (DAIS), Ca' Foscari University of Venice,
Venice, Italy

Abstract. The identification of emergent structures in dynamical systems is a major challenge in complex systems science. In particular, the formation of intermediate-level dynamical structures is of particular interest for what concerns biological as well as artificial systems. In this work, we present a set of measures aimed at identifying groups of elements that behave in a coherent and coordinated way and that loosely interact with the rest of the system (the so-called "relevant sets"). These measures are based on Shannon entropy, and they are an extension of a measure introduced for detecting clusters in biological neural networks. Even if our results are still preliminary, we have evidence for showing that our approach is able to identify and partially characterise the relevant sets in some artificial systems, and that this way is more powerful than usual measures based on statistical correlation. In this work, the two measures that contribute to the cluster index, previously adopted in the analysis of neural networks, i.e. integration and mutual information, are analysed separately in order to enhance the overall performance of the so-called dynamical cluster index. Although this latter variable already provides useful information about highly integrated subsystems, the analysis of the different parts of the index are extremely useful to better characterise the nature of the sub-systems.

1 Introduction

We have recently shown that the cluster index (CI), a measure that had been introduced by Tononi and Edelman [3,4] to study the behaviour of neural networks close to their stationary states, can be suitably generalised and profitably applied to a range of different systems, in order to identify subsets of the system variables that change in time in a coordinated way. In particular, our *dynamical*

© Springer International Publishing Switzerland 2014
C. Pizzuti and G. Spezzano (Eds.): WIVACE 2014, CCIS 445, pp. 140–150, 2014.
DOI: 10.1007/978-3-319-12745-3_12

cluster index (DCI) has been successfully applied to artificial boolean networks, to models of catalytic reaction networks and of the genetic network of a living organism and to robot control [2,5,6]. Precise definitions of CI and DCI are provided in Sect. 2.

The study of these subsets is particularly important because they may deeply affect the organisation and the dynamics of the whole system and may represent a clear example of so-called "sandwiched emergence" [1]. For ease of notation, we will often refer to the subsets of variables that change in a coordinated way as "relevant subsets"(RS) of the whole system.[1]

There are of course several methods to identify clusters of nodes in a network, based upon the existing links, but the DCI can be applied also without any prior knowledge on the interactions among the relevant variables. This property is shared also by correlation methods; with respect to these latter ones, the advantage of the DCI is that it is not limited to binary relations but it can be applied to clusters of any size, although the computational load scales steeply with the number of elements.

The basic idea is fairly simple, but there are some subtleties that deserve a better understanding. The main goal of this paper is to highlight and analyse some of these aspects. As we shall see, these studies will also lead us to enlarging the set of measures that can be profitably used.

The dynamical cluster index can be used on an arbitrary series of variable values taken at equal times, e.g. a time series of the values of the variables associated to the various nodes of a network. In the case of autonomous systems, an important issue concerns which kind of information on the relevant subsets can be obtained by analysing asymptotic values (i.e. attractors) vs. transients. This important aspect is investigated in Sect. 5.

Section 2 summarises the CI and the DCI, along with a discussion on the main issues concerning its estimation and application in Sect. 3. In Sects. 4 and 5 we illustrate the results achieved in the assessment and application of the DCI on discrete dynamical systems. The paper concludes in Sect. 6 with an outlook to future work and open problems.

2   The Dynamical Cluster Index

Our aim is to identify subsets of elements in a system with a high interaction among themselves, i.e., high *integration*, with respect to the information exchanged with the overall system.

In order to describe the dynamical cluster index, let us consider a system U, composed of N elements assuming finite and discrete values. The cluster index is an information theoretical measure based on the Shannon entropy of both the single elements and sets of elements in U.

According to the information theory, the entropy of an element x_i is defined as:

$$H(x_i) = - \sum_{v \in V_i} p(v) \ log \ p(v) \tag{1}$$

---

[1] This also to avoid confusion with the term *cluster*, often used with a slightly different meaning in the context of data mining.

where V_i is the set of the possible values of x_i and $p(v)$ the probability of occurrence of symbol v. With regard to the entropy of the subsets belonging to U, given that the entropy of a pair of elements x_i and x_j is defined by means of their joint probabilities:

$$H(x_i, x_j) = -\sum_{v \in V_i} \sum_{w \in V_j} p(v, w) \ log \ p(v, w) \tag{2}$$

it is worthwhile to notice that Eq. 2 can be extended to sets of k elements simply considering the probability of occurrence of vectors of k values.

Since we deal with a dynamic defined by a series of discrete states of an element, probabilities can be approximated by means of the frequencies of the observed values.

Accordingly, since the only information we need concerns the collection of the states of the elements in the system used to compute the frequencies of the possible states, we do not rely on the exact sequence of these states.

In accordance with the aforementioned definition of highly *integrated* subsystems exchanging information with the rest of the system, the cluster index $C(S)$ of a set S of k elements is defined as the ratio between the *integration* $I(S)$ and the *mutual information* $M(S; U - S)$ between S and the rest of the system $U - S$. Let us define the integration as:

$$I(S) = \sum_{x \in S} H(x) - H(S). \tag{3}$$

$I(S)$ basically represents the deviation from statistical independence of the k elements in S. Then, let us define the mutual information $M(S; U - S)$ as:

$$M(S; U - S) \equiv H(S) + H(S|U - S) = H(S) + H(U - S) - H(S, U - S) \tag{4}$$

where $H(A|B)$ is the conditional entropy and $H(A, B)$ the joint entropy. Finally, the cluster index $C(S)$ is defined by:

$$C(S) = \frac{I(S)}{M(S; U - S)}. \tag{5}$$

Since the joint entropy cannot assume values greater than the sum of the single entropies, both the indices are not negative. Then, according to the nature of the index, particular attention must be devoted to some peculiar values of the measures composing the index, such as $I = 0 \wedge M \neq 0 \Rightarrow C(S) = 0$, $M = 0 \Rightarrow C(S)$ not defined and $I > 0 \wedge M = 0 \Rightarrow C(S) = \infty$.

The first case indicates no integration at all, the second case stands for lack of information and the third case accounts for a subset S statistically independent from the rest of the system, i.e., not exchanging information with respect to the observed variable.

With regard to the properties of the cluster index, $C(S)$ scales with S, hence subsystems of different size cannot be directly compared. To overcome this limitation, cluster indices of different dimension need to be normalised. To this aim, let us define a reference system, i.e., the homogeneous system U_h, randomly generated in accordance with the probability of each single state measured in the original system U, along all its series of states. Then, in order to have a reference value for the cluster index for each dimension, for each subsystem size of U_h the average integration $\langle I_h \rangle$ and the average mutual information $\langle M_h \rangle$ are computed.

Hence, the cluster index value of any subsystem S can be normalised by means of the appropriate normalisation constant based on the size S:

$$C'(S) = \frac{I(S)}{\langle I_h \rangle} \Big/ \frac{M(S; U - S)}{\langle M_h \rangle} \tag{6}$$

In order to compare the cluster indices computed on all the assessed clusters, a statistical significance index T_c is computed by normalising both the cluster indices of the analysed system and the homogeneous one:

$$T_c(S) = \frac{C'(S) - \langle C'_h \rangle}{\sigma(C'_h)} = \frac{\nu \cdot C - \nu \cdot \langle C_h \rangle}{\nu \cdot \sigma(C_h)} = \frac{C(S) - \langle C_h \rangle}{\sigma(C_h)} \tag{7}$$

where $\langle C'_h \rangle$ and $\sigma(C'_h)$ stand for the average and the standard deviation of the population of normalised cluster indices with the same size of S from the homogeneous system and $\nu = \frac{\langle M I_h \rangle}{\langle I_h \rangle}$ stands for the normalisation constant. Note that in Eq. 7 we prove that, in order to compute the statistical significance, the normalisation with respect ν is not a necessary step. Finally, T_c is adopted to rank the evaluated clusters.

Since the cluster index will be applied in the following to dynamical systems, we will refer to this measure as *Dynamical cluster index*.

3   DCI: Remarks and Observations

As previously stated, the two factors used to compute the DCI of a candidate RS are the integration of its elements and the mutual information between the subset and the rest of the system. Therefore, the DCI of a subset S of variables in a system—i.e., a possible candidate RS—is then estimated by collecting a set of system states and then computing the entropy values of the possible combinations of variables in S. Therefore, one just needs a collection of observations of the system and, in principle, no information on the topology or the internal mechanics of the system are needed.

The rationale behind the ratio between I and MI is that a candidate RS should express a dynamics more intense among its components than with the rest of the system. In spite of the simplicity of the concept, one must carefully consider the meaning of the two quantities and the way they are combined. In fact, the implicit assumption here is that it holds $MI > 0$ and that the orders of magnitude of I and MI are reasonably close. However, there might be cases in which $MI \approx 0 \land I > 0$, which denotes integrated subsets dynamically independent from the rest of the system. Furthermore, we may be interested in finding the most integrated subsystems among the ones that exchange information with the rest of the system. Moreover, it is important to remark that insightful information can be also provided by analysing the two factors composing the DCI, before computing their ratio. For this reason, an assessment of the information brought by the individual values of I and MI can shed light on the potential of the application of these information-theoretic measures to detect candidate RSs.

4   Integration vs. Mutual Information

To assess the contribution of I and MI we first studied their statistical behaviour with respect to subsystem size. We analysed several dynamical systems and the results obtained show that the general trend respects the theoretical findings [4], as shown in Figs. 1 and 2.

As we can observe, I scales approximately linearly, whilst MI is non-monotonic. It can be proven that the maximal value of I for a subset of size n equals $n - 1$. Accordingly we can define a rescaled version of I, rI, useful to compare the integration of subsets of different dimensions.

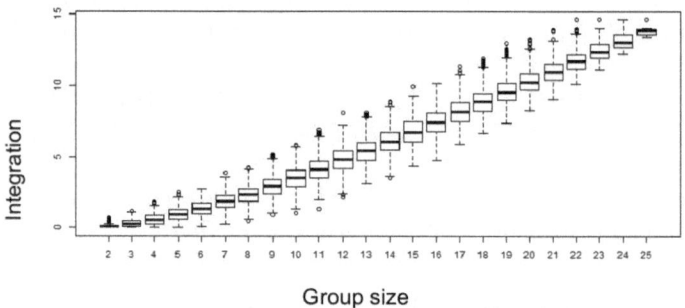

Fig. 1. Typical statistical behaviour of integration for different subsystem sizes. The system analysed is a catalytic reaction system [6] composed of 26 molecular species.

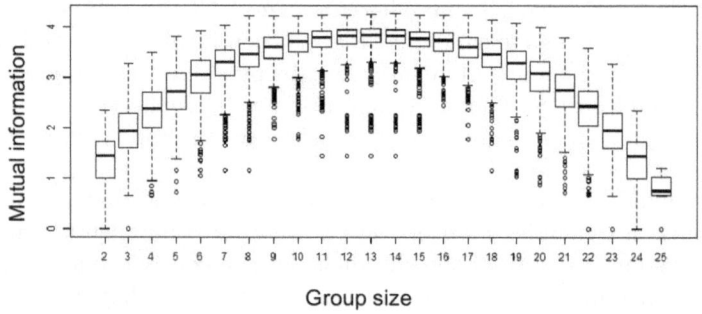

Fig. 2. Typical statistical behaviour of mutual information for different subsystem sizes. The system analysed is a catalytic reaction system [6] composed of 26 molecular species.

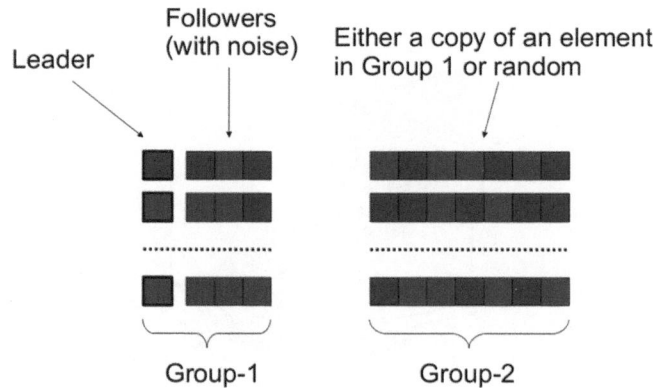

Fig. 3. Data generated according to a simple leader-followers model.

Moreover, the addition of a variable x to the considered subset S increases I by 1 if x is depends deterministically on any variable in S, while it leaves I unchanged if it assumes random values. These properties may be useful to reckon the relative importance of integration values computed for subsystems of different size.

4.1 Experiments on a Tuneable Model

As discussed in Sect. 3, we are interested in understanding the individual informative contribution of I and MI. To this aim, we studied these measures on a simple model in which the integration among variables in a subsystem under observation and its mutual information can be tuned by acting on two parameters. The model abstracts from specific functional relations among elements of the system and could resemble a basic leader-followers model (see Fig. 3). The system is composed of a vector of n binary variables $x_1, x_2, \ldots, x_n$, e.g., representing the opinion in favour or against a given proposal. The model generates independent observations of the system state, i.e., each observation is a binary n-vector generated independently of the others, on the basis of the following rules:

- Variables are divided into two groups, $G_1 = [x_1, \ldots, x_k]$ and $G_2 = [x_{k+1}, \ldots, x_n]$;
- x_1 is called the *leader* and it is assigned a random value in $\{0, 1\}$;
- the value of the *followers* $x_2, \ldots, x_k$ is set as a copy of x_1 with probability $1 - p_{noise}$ and randomly with probability p_{noise};
- the values of elements of G_2 are assigned as a copy of a random element in G_1 with probability p_{copy}, or a random value with probability $1 - p_{copy}$.

It is possible to tune the integration among elements in G_1 and the mutual information between G_1 and G_2 by changing p_{noise} and p_{copy}. Note that, given significant level of integration, we have two notable cases:

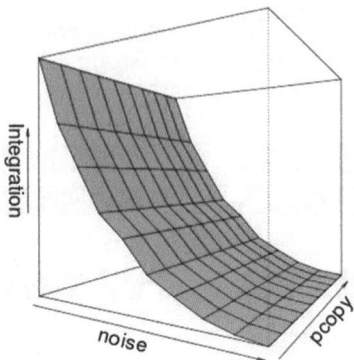

Fig. 4. Integration of G_1 as a function of p_{noise} and p_{copy}.

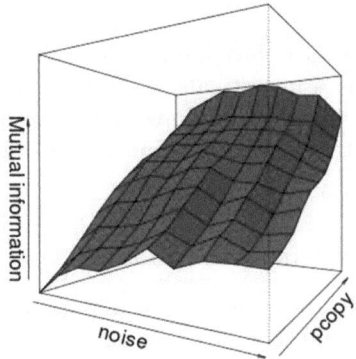

Fig. 5. Mutual information between G_1 and G_2 as a function of p_{noise} and p_{copy}.

(a) $MI \approx 0 \rightarrow$ isolated (possibly integrated) RS;
(b) $MI \gg 0 \rightarrow$ integrated and segregated cluster.

The possible scenarios which can be obtained by tuning p_{noise} and p_{copy} can be conveniently illustrated by a 3-dimensional plot. Figure 4 shows the behaviour of integration of G_1 as a function of p_{noise} and p_{copy}. We can observe that it is a decreasing function of p_{noise}, while it is independent of p_{copy} (by definition, indeed). The behaviour of the mutual information between G_1 and G_2 is depicted in Fig. 5. As we can observe, MI increases fast with p_{copy}, as this parameters increases the correlation between variables in G_2 and G_1. Moreover, it also increases with p_{noise}, but the reason is that the correlation among variables in G_1 increases the randomness of variables in G_1, which than behaves similarly to the variables in G_2.

The case of $MI \approx 0$ (case (a)), corresponds to the situation in which G_1 is almost completely independent of G_2 and can be easily detected by observing only MI. Conversely, if we are interested in discovering G_1 as significant RS (case *(b))*, then we would consider cases in which both I and MI are significantly high. In our experiments, these cases correspond to $p_{noise} < 0.3$ and $0 < p_{copy} < 0.5$. In these scenarios, we found that it is possible to detect G_1 by the sole use of I, divided by $n - 1$. It is important to mention that, when p_{noise} is slightly higher, it is necessary to resort to the computation of the DCI to detect group G_1. This result brings evidence to the observation that the method based on the cluster index is superior to usual correlation methods and hierarchical clustering.

4.2 Combined Use of I and MI

Integration and mutual information provide complimentary pieces of information which can be combined observing the relation MI vs. rI. We analysed this relation in systems we studied in previous works and we discovered that the re-scaling of I makes it possible to detect candidate RSs without the need of performing the whole computation of the DCI.

Figure 6 shows a typical case, corresponding to the analysis of a boolean network obtained by artificial evolution to control for a robot that has to perform obstacle avoidance [2]. As we can observe the points mostly follow the MI non-monotonic curve, but there are some outliers at high values of integration and significant values of mutual information (diamonds at the bottom-right corner of the figure). After a direct inspection of the network, we observed that these points correspond to couples or triplets of nodes including one of the proximity sensors of the robot and other nodes either involved in the robot motion—i.e.,

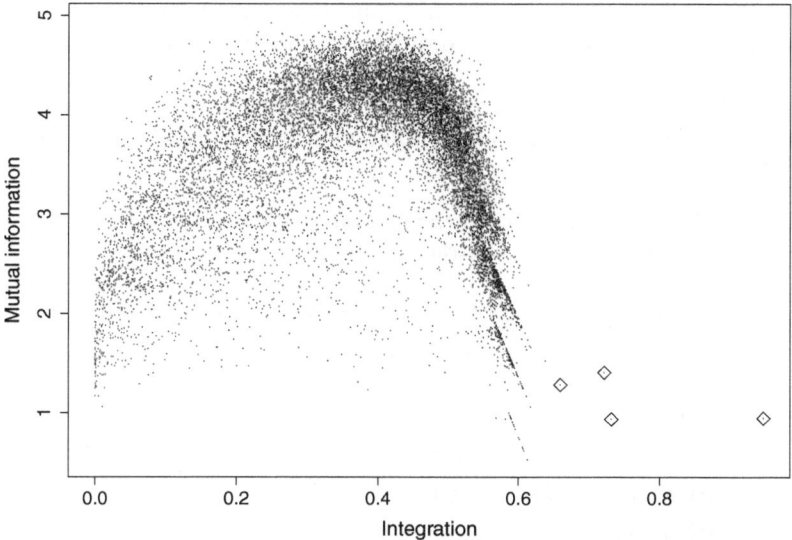

Fig. 6. Mutual information vs. rescaled integration $rI = I/(n-1)$ of a boolean network used to control a robot. Diamonds denote the RSs with highest values of integration and significant levels of mutual information.

connected to a wheel—or simply ruled by a dynamics dependent from the sensor. We do not discuss here the details and the consequences of this analysis for the case at hand, but we rather emphasise that the combined use of rI and MI makes it possible to easily identify candidate RSs.

5   Analysis of the Information Provided by Transients and Attractors in Dynamical Systems

Let us consider now the nature of the information gathered from the analysis of dynamical systems with respect to the dynamical state of the system when data are collected, during the transient or within an attractor. Should the results be somehow comparable, then we would have a much more flexible experimental setting, which could be also applied to situations in which system's attractors and their basins are difficult to be estimated.

In order to prevent the presence of the aforementioned peculiar values of the DCI, the subsets are ranked according to the rI index. So, we computed the rI on the data composed of the system's states collected during the transients (from random initial states). In Table 1 we summarise the outcome of the comparison on some typical systems.[2] The first 30 positions in the rank are then compared between the two cases and the common candidate RSs are counted (2nd column of Table 1). As we can see, in some cases the match is quite high. However, by direct inspection of the data, we observed that in all cases there is a relatively large gap between the first positions and the others. Therefore, a more accurate comparison could be done by looking at the whole rI distribution and take into consideration only the positions higher than this gap. As a particular case, we can note that all the top ranked candidate RSs in transients and on attractors are the same.

Transients and attractors have different characteristics, at least because during transients there is a continuous exchange of information within the system:

Table 1. Fraction of matches among the first 30 candidate RSs w.r.t. rI between the case with data series composed of attractors and with transients.

System	# of matches	Match on 1st position
Arabidopsis	3/30	no
BN1	30/30	yes
BN2	29/30	yes
BN3	10/30	yes
RBN, $n = 20, k = 2$	6/30	yes
BN-robot	13/30	yes

[2] For Arabidopsis, BN1, BN2, and BN3 see [6], while for BN-robot see [2]. RBN is a randomly generated boolean network with n nodes and k inputs per node.

Table 2. Comparison of the occurrences of cases with $MI = 0$.

System	On attractors		On transients		On attractors $\geq$ On transients	
	$I \neq 0$	$I = 0$	$I \neq 0$	$I = 0$	$I \neq 0$	$I = 0$
Arabidopsis	10	69	3	39	yes	yes
BN1	0	0	0	0	yes	yes
BN2	0	6	0	6	yes	yes
BN3	0	27	0	0	yes	yes
RBN, $n = 20, k = 2$	106	788	0	0	yes	yes
BN-robot	0	0	0	0	yes	yes

this exchange should exceed the boundaries of the identified RSs, because not all the elements of the systems are in equilibrium. Indeed, it is possible to detect this exchange process by considering the number of identified RSs that have MI equal to zero. This number is always higher on attractors than on transient, both considering RSs that have I different or equal to zero (see Table 2).

More data are required to formulate a solid statement, but we have anyway some evidence to support the conjecture that the indices presented here are sufficiently robust against the data series collected about the system at hand, especially for the top ranked candidate RSs. This result is promising, as it suggests that the information collected during transients and on the attractors leads to similar results with respect to the frequency of the states observed and it also makes it possible to extend the range of the possible applications of DCI to systems that never reach a true attractor.

6   Conclusion

In this work we have presented preliminary results of the study of the measures involved in the computation of the DCI, namely integration and mutual information. We have shown that, besides providing notable results if combined as a ratio, they also bring valuable pieces of information on the candidate RSs. We have discussed the role and contribution of both the measures and characterised the meaning of their extreme values, such as the cases with $MI \approx 0$ and $MI \gg 0$. In addition, we have presented results supporting the conjecture that rescaled integration is particularly effective in detecting RSs, also in the cases in which $MI \neq 0$.

Moreover, we have also investigated the robustness of the results attained by collecting the states of a dynamical system on the attractors or on the transients. Results show that the top ranked positions according to rI have a good match between the two cases, suggesting that the measure is quite robust against the procedure of collection of states. Furthermore, an analysis of the information exchanged between the RSs and the rest of the system show that during transients the amount of information transferred is higher than that on attractors.

In future work we plan to extend this analysis to a larger number of systems, so as to bring more evidence on the conjectures formulated in this paper.

We also plan to apply the method to the analysis of real networks, trying to combine static information—e.g., topological—and dynamic information.

Acknowledgements. The research leading to these results has received funding from the European Union Seventh Framework Programme (FP7/2007-2013) under grant agreement n. 284625.

References

1. Lane, D., van der Leeuw, S., Pumain, D., West, G. (eds.): Complexity Perspectives on Innovation and Social Change, vol. 7. Springer, Heidelberg (2009)
2. Roli, A., Villani, M., Serra, R., Garattoni, L., Pinciroli, C., Birattari, M.: Identification of dynamical structures in artificial brains: an analysis of boolean network controlled robots. In: Baldoni, M., Baroglio, C., Boella, G., Micalizio, R. (eds.) AI*IA 2013. LNCS, vol. 8249, pp. 324–335. Springer, Heidelberg (2013)
3. Sporns, O., Tononi, G., Edelman, G.M.: Connectivity and complexity: the relationship between neuroanatomy and brain dynamics. Neural Netw. Official J. Int. Neural Netw. Soc. **13**(8–9), 909–922 (2000)
4. Tononi, G., McIntosh, A., Russel, D., Edelman, G.: Functional clustering: identifying strongly interactive brain regions in neuroimaging data. Neuroimage **7**, 133–149 (1998)
5. Villani, M., Benedettini, S., Roli, A., Lane, D., Poli, I., Serra, R.: Identifying emergent dynamical structures in network models. In: Proceedings of WIRN 2013 – Italian Workshop on Neural Networks (2013)
6. Villani, M., Filisetti, A., Benedettini, S., Roli, A., Lane, D., Serra, R.: The detection of intermediate-level emergent structures and patterns. In: Liò, P., Miglino, O., Nicosia, G., Nolfi, S., Pavone, M. (eds.) ECAL 2013, pp. 372–378. The MIT Press, Cambridge (2013)

Investigating the Role of Network Topology and Dynamical Regimes on the Dynamics of a Cell Differentiation Model

Alex Graudenzi[1][(✉)], Chiara Damiani[1,2][(✉)], Andrea Paroni[1],
Alessandro Filisetti[3], Marco Villani[3,4], Roberto Serra[3,4],
and Marco Antoniotti[1]

[1] Department of Informatics, Systems and Communication,
University of Milan-Bicocca, Milan, Italy
{alex.graudenzi,chiara.damiani}@unimib.it
[2] SYSBIO - Centre for Systems Biology, Piazza Della Scienza 2, 20126 Milan, Italy
[3] European Centre for Living Technology,
University Ca' Foscari of Venice, Venice, Italy
[4] Department of Physics, Informatics and Mathematics,
University of Modena and Reggio Emilia, Modena, Italy

Abstract. The characterization of the generic properties underlying the complex interplay ruling cell differentiation is one of the goals of modern biology. To this end, we rely on a powerful and general dynamical model of cell differentiation, which defines differentiation hierarchies on the basis of the stability of gene activation patterns against biological noise.

In particular, in this work we investigate the role of the topology (i.e. scale-free or random) and of the dynamical regime (i.e. ordered, critical or disordered) of gene regulatory networks on the model dynamics. Two real lineage commitment trees, i.e. intestinal crypts and hematopoietic cells, are compared with the hierarchies emerging from the dynamics of ensembles of randomly simulated networks.

Briefly, critical networks with random topology seem to display a wider range of possible behaviours as compared to the others, hence suggesting an intrinsic dynamical heterogeneity that may be fundamental in defining different differentiation trees. Conversely, scale-free networks show a generally more ordered dynamics, which limit the overall variability, yet containing the effect of possible genomic perturbations. Interestingly, a considerable number of networks across all types show emergent trees that are biologically plausible, suggesting that a relatively wide portion of the networks space may be suitable, without the need for a fine tuning of the parameters.

1 Introduction

In the last years a *dynamical* interpretation of gene regulation has gained an always greater attention, as a complement to the *static* consideration of the interactions among genomic entities, as classically done through the *systems-biology*

© Springer International Publishing Switzerland 2014
C. Pizzuti and G. Spezzano (Eds.): WIVACE 2014, CCIS 445, pp. 151–168, 2014.
DOI: 10.1007/978-3-319-12745-3_13

approach [17]. The focus switches from the static description of the involved entities and interactions to the analysis of the *emergent* collective behaviors, that are the *gene activation patterns* characterizing the different phenotypic functions of cells, such as cell types or modes. Each organism is, in fact, characterized by a unique gene regulatory network, but is able to exhibit a broad range of distinct phenotypes. Any (phenotypic) function is determined by a specific gene activation pattern, which is the result of the joint dynamical interaction among genes.

Accordingly, *gene regulatory networks* (GRNs) can be modeled as *dynamical systems* and the focus is on the analysis of their "*attractors*" [14,21,30][1]. Many different mathematical and computational models aimed at representing the dynamics of GRNs have been developed through the years, with different goals and applications and most of them are rooted in complex systems science and statistical physics (see, e.g., [38] for a recent review).

In particular, we here present a study regarding a dynamical model of cell differentiation[2], introduced in [39] and based on *Noisy Random Boolean Networks* (NRBNs, [27,31]). This simplified model of gene regulation considers genes as simply active/inactive and describes simplified regulatory interactions (not explicitly considering the underlying biochemical machinery), focusing on the dynamical behaviour emerging from the interaction of the genes. In line with the complex systems approach the major goal is to investigate the *generic* properties of gene networks, that are those properties shared by a broad range of systems and organisms. To this end, a powerful methodological means is that provided by the statistical analysis of *ensembles* of randomly simulated networks with specific features, in order to scan the huge space in which real networks, on which complete information is still missing, might be found. Despite the underlying abstractions, this modeling approach was proven fruitful in describing properties of real networks, as in, e.g.,[21,22,29,33,34,36].

The model of cell differentiation we here analyze is abstract and general, i.e. it is not referred to any specific system or organism and it is based on two specific dynamical properties of the attractors of GRNs: their *robustness* against random or selective perturbations (i.e. genomic mutations/alterations) and their *reachability*, defined as the likelihood of observing a certain attractor. Starting from the underlying hypotheses that biological noise and specific perturbations can trigger transitions among the GRN attractors (i.e. the gene activation patterns) and that higher level of noise were detected in less differentiated cells [8,10,11,13,24],

---

[1] Given a GRN dynamical model, the long term evolution will confine the cellular states in a specific region of the state space, i.e. an *attractor*, in which the values of the variables can be fixed over time, can be characterized by oscillatory periodic regimes, or even by more particular non-periodic dynamics (in non finite-states deterministic models). Any GRN can be characterized by the presence of different attractors, reachable from distinct initial conditions. The attractors represent coherent activation patterns of genes.

[2] Cell differentiation is the process according to which the progeny of stem cells progressively develops into different and always more specialized cell types, crossing various intermediate stages.

two constraints must be respected: (*i*) the attractors of a specific cell type must be sufficiently robust against perturbations not to compromise the functioning of the cell, (*ii*) in less differentiated cell types attractors must be sufficiently sensitive to perturbations to account for different cell fates when differentiating. These constraints are satisfied in our model by assuming that: (*i*) cell types or modes are characterized by sets of attractors in which the GRN can wander via noise or specific signals; (*ii*) the process of progressive specialization is characterized by a gradual improvement of the noise-control mechanisms that hinder the transitions, from less differentiated cells, sensitive to noise, to fully differentiated stages, robust against noise. In this way it is possible to define a hierarchy connecting the differentiation levels and cell fate decisions are then driven by stochastic fluctuations or triggered by specific signals.

Many important features of the differentiation process are reproduced by the model, such as: (*i*) the presence of different degrees of differentiation, that span from totipotent stem cells to fully differentiated cells in a well-defined hierarchy, which determines a lineage tree; (*ii*) the *stochastic differentiation*, according to which populations of identical multipotent cells stochastically generate different cell types; (*iii*) the *deterministic differentiation*, in which specific signals trigger the progress of multipotent cells into more differentiated types, in well-specified lineages; (*iv*) the *limited reversibility* in which, under the action of appropriate signals, the cell can revert its lineage specification; (*v*) the *induced pluripotency*, according to which fully differentiated cells can come back to a pluripotent state by modifying the expression of specific genes [41]; (*vi*) the induced change of cell type, in which the modification of the expression of few genes can directly convert one differentiated cell type into another.

The emerging lineage commitment tree can be matched against real differentiation trees as those in Fig. 1, through simple tree-matching algorithms. The analysis of a large number of GRNs matching/non-matching real trees can then provide some cues for explaining the properties ruling the complex differentiation interplay.

Thus, the focus of this work is to analyze the influence of: (*i*) the *topology* and of (*ii*) the *dynamical regime* of the GRNs on the key dynamical properties and especially on their robustness and reachability, with particular regard to the emerging differentiation hierarchy. In particular, two real differentiation trees will be targeted by our model, *intestinal crypts* and *hematopoietic cells*.

In regard to the former analysis, classical studies on RBNs and NRBNs involve random topologies (i.e. Erdos-Renyi-based [9]), which determine a Poissonian distribution of the connection. Even though present data still do not allow to draw definitive conclusion on the topology underlying real GRNs, it is sound to investigate the relation that different kinds of topologies may have on the overall emerging behaviour. In particular, *scale-free* networks [3] have raised a considerable interest and were shown to approximate several networks, including metabolic and protein networks[3]. Therefore, we here present the analysis of a study comparing the behaviour of NRBNs with random and scale-free topologies.

---

[3] Excluding the high degree exponential cutoffs due to the limited size of the networks.

The second analysis concerns the influence that the dynamical regimes have on the properties of the GRNs relevant to the differentiation model. In short, the dynamical regime of a RBN is defined on the basis of the sensitivity to the initial conditions, that is the response to small perturbations. If small perturbations tend to lead to different attractors, i.e. the perturbation propagates, the network can be considered as *disorderered* (sometimes refered to as *chaotic*) and vice versa. Networks characterized by disordered regimes are (on the average) characterized by a larger number of longer attractors and vice versa. It was analytically proved that by varying certain key structural parameters of RBNs it is possible to switch across the dynamical regimes. A very interesting dynamical region is at the boundary between disordered and ordered phase and is defined as *"critical"*, sometimes referred to as the *"edge of chaos"* [23]. It was hypothesized the biological systems, and GRNs in particular, may live and evolve in this specific region, which would allow an optimal trade-off between robustness and evolvibility [21]. Experimental evidences in support of this hypothesis are provided, e.g., in [29,33,36]. In this work we aim at investigating how the dynamical regime can influence the properties of the cell differentiation model, with specific regard to the suitability for a matching with real differentiation trees.

As specified above, the two specific biological systems that were chosen as test beds for the model are intestinal crypts and hematopoietic cells. Intestinal crypts are invaginations in the intestine connective tissue in which tumors are supposed to originate from some partially known gene and pathway alterations affecting the stem cell niche [1]. A complex differentiation process rules the overall homeostasis, in terms of stratification of cell populations of distinct types, coordinate migration, dynamic turnover, etc. Conversely, in hematopoietic cells, the differentiation can be interpreted as a trajectory among attractors, involving the transcriptome as the state space of cell populations and the miRNome as tuning mechanism, and that starts from multipotent hematopoietic stem cells giving rise to a hierarchy of progenitor populations with more limited lineage potential, eventually leading to mature blood cell types [Felli et al. 2010].

In both cases we are interested in comparing the already mapped lineage trees (Fig. 1) with the trees emerging from the dynamics of randomly generated networks. In this way it is possible to investigate the structural and dynamical features of the suitable networks, possibly providing hypotheses on the generic properties of real networks.

This study is a part of a series of articles aimed at the analysis of the properties of the model of cell differentiation introduced in [39]. Other models have been proposed in course of the time, even if with distinct modeling approaches, e.g., [15,17,26].

In Sect. 2 the dynamical model of cell differentiation is described. In Sect. 3 the analyses concerning the topology, the dynamical regime and the lineage tree comparison are described. Section 4 the results of the simulations are shows, whereas Sect. 5 contains the final remarks.

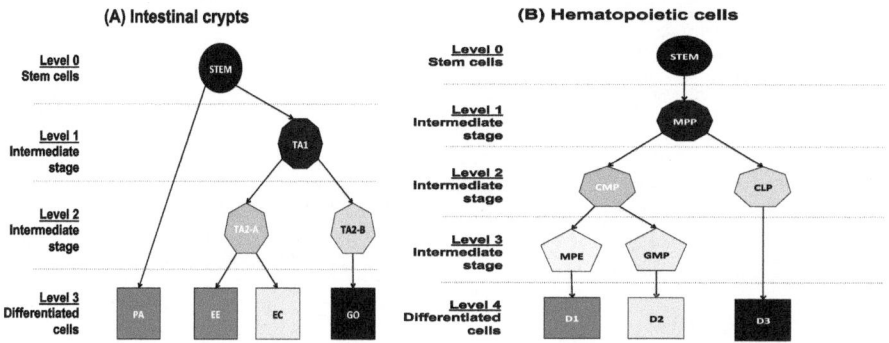

Fig. 1. Differentiation trees in intestinal crypts and hematopoietic cells. In **(A)** the crypt differentiation tree is shown, involving stem, transit amplifying stage (**TA1, TA2-A, TA2-B**), Paneth (**PA**), Goblet (**GO**), enteroendocrine (**EE**) and enterocyte (**EC**) cells [6]. In **(B)** the hematopoietic differentiation tree is shown, involving stem, multipotent progenitors (**MPP**), common myeloid progenitors (**CMP**), common lymphoid progenitors (**CLP**), megakaryocyte-erythroid progenitors (**MEP**), granulocyte-macrophage progenitors (**GMP**). The finally differentiate cells are considered only as distinct classes of differentiated cells (**D1, D2, D3**) [40].

2   A Dynamical Model of Cell Differentiation

Here we will briefly outline the main features of the dynamical model of cell differentiation introduced in [39], for a more exhaustive description please refer to the original work.

Random Boolean Networks (RBNs, [19–21]) are an abstract model of *Gene Regulatory Network* (GRN). These networks are directed graphs whose nodes are binary variables x_i that model the activation/inactivation of the associated gene (i.e. production of a specific protein or RNA); the edges symbolize the regulatory paths. A Boolean updating function f_i is associated to each x_i and the update occurs synchronously at discrete time step for each node of the network, according to the value of the inputs nodes at the previous time step. So, if the state at time t of a RBN is the binary vector $\mathbf{x}(t)$, the i-th component of state $\mathbf{x}(t+1)$ is:

$$\mathbf{x}(t+1)_i \stackrel{\text{def}}{=} f_i(\mathbf{x}(t)) \tag{1}$$

Since the state-space is finite (i.e. there exist at most 2^n vectors in $\{0,1\}^n$) and the dynamics is fully deterministic, the system will end up in a limit cycle from any initial condition $\mathbf{x}(0)$[4]. Such a cycle is an *attractor* of the RBN and the sequence of states from $\mathbf{x}(0)$ to the cycle is the *transient* of the attractor; accordingly, the set of initial conditions ending up to the same attractor is named *basin of attraction*. Notice that attractors correspond then to gene activation patterns.

---

[4] We here use the so called *quenched* model [21], in which both the graph and the boolean functions do not change in time.

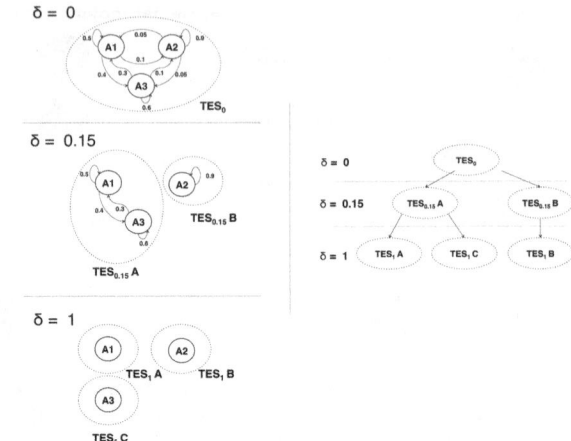

Fig. 2. Example ATNs and TESs. An example of the threshold-dependent ATN and the corresponding tree-like TES landscape. The circle nodes are attractors of an example NRBN, the edges represent the relative frequency of transitions from one attractor to another one, after a 1 time step-flip of a random node in a random state of the attractor (performed an elevated number of times). In this case we show three different values of threshold, i.e.: $\delta = 0$, $\delta = 0.15$ and $\delta = 1$. TESs, i.e. strongly connected components in the threshold-dependent ATN with no outgoing links are represented through dotted lines and the relative threshold is indicated in the subscripted index. In the right diagram it is shown the tree-like representation of the TES landscape.

Considering that biological noise is known to play a role in several key cellular processes and in differentiation as well [4, 8, 12, 16, 18, 24, 25, 28, 37], an extension of RBNs accounting for stochastic fluctuations was developed, named Noisy Random Boolean Networks, NRBN [27, 31].

The key notion in the NRBN model, as presented in [39], is the definition of an *attractor transition network* (ATN). The ATN is a stability matrix whose entries represent the probability of switching among RBN attractors, as a consequence or random flips of the value of the nodes (i.e. $x_i = 1$ gets flipped to 0, and vice versa) in an attractor. From another perspective, the ATN describes the possibility of wandering among the attractors as a consequence of the smallest perturbation which can affect a RBN, i.e. the flip. In [31] we showed that the transition probability among attractors decreases sub-linearly with respect to the graph size.

The definition of the ATN for a certain NRBN allows to represent the phenomenon of hierarchical stochastic differentiation. The underlying biological idea is the following: any cell type is characterize by a certain number of gene activation patterns in which the cell can wander as a consequence of random fluctuations and specific signals; besides, more differentiated cells are characterized by a better robustness against noise, because of more refined noise-control mechanism and, consequently, can roam in smaller portion of the state space, as experimentally proven by the higher level of noise in gene activation that has been detected in undifferentiated cells [10, 11, 13, 24].

In our model it possible to introduce a *threshold* $\delta \in [0,1]$ that is used to remove the transitions that are too rare, i.e. that cannot occur within the life-time time of a cell, so defining a *threshold dependent-ATN*. Intuitively, higher thresholds represent better noise-control mechanism and are associated to progressively more differentiated cells. Each cell type of a cell is then associated to a so-called Threshold Ergodic Set (TES), as defined in [39]: given a certain *threshold* $\delta \in [0,1]$ a TES_δ is:

- a *strongly connected component*, SCC, in a threshold dependent-ATN,
- in which there are no outgoing links from any attractors belonging to the SCC toward an external attractor

TESs represent coherent stable ways of functioning of the same genome even in the presence of noise, i.e. cell types or modes. In this way, toti-/pluri-potent stem cells, for instance, will be characterized by a very low values of the threshold and will wander across various gene patterns, in order to resemble their biological capability of differentiating in various cellular types. Conversely, differentiated types will be associated to high thresholds. More in general, different thresholds will be associated to different degrees of differentiation. The phenomenon of stochastic differentiation is then defined as follows: the fate of a cell depends on the attractor where the cell is when the cell divides and the threshold increases. The new cell type will be that corresponding to the TES to which the attractor belongs in correspondence of the new threshold level. This allow to define parent and children types of cells and, accordingly, to draw a coherent a hierarchical differentiation tree. In principle, it is possible to map any desired differentiation tree to a *partial order* over thresholds (see Fig. 2).

Notice that in this case we intend to model only the stochastic differentiation process, whereas the model is capable of representing the deterministic differentiation as well[5]. Future analyses will investigate the repercussions of a change in the topology and in the dynamical regimes also with respect to it.

3 The Influence of Topology and of the Dynamical Regimes

The goal of the current work is to disentangle the effects of (*i*) the topology and of (*ii*) the dynamical regime of a NRBN on the properties of the TESs and, accordingly of the emerging differentiation tree. Considering that the information on real GRNs is still partial, along the lines of the complex systems approach we here focus on the generation of large ensembles of randomly generated networks with specific structural feature, in order to investigate the emergent dynamical properties of classes of networks and to relate them to the differentiation process.

---

[5] In several processes, e.g., during the embryogenesis, cell differentiation is not stochastic but it is driven towards precise, repeatable types by specific chemical signals. In our model, it was shown that certain genes, called *switch genes*, if permanently perturbed and coupled with a change in the threshold always leads the system through the same differentiation pathway. In other words, nodes that uniquely determine to which TES the system will evolve, i.e. deterministic differentiation.

Topological classes: random vs. scale-free. It was recently shown that genetic and metabolic networks might have a different topology than the Erdos-Renyi [9] one and, in particular, the presence of a large number of scarcely connected nodes and of a small number of largely connected nodes (i.e. the *hubs*) hints at the possible presence of an underlying power law in the distribution of the connections [3]. If $P(k)$ is the probability for a particular node of having k connections, the power law is defined as:

$$P(k) = \frac{1}{Z}k^{-\gamma} \tag{2}$$

$$Z = \sum_{i=1}^{k_{max}} i^{-\gamma} \tag{3}$$

where k can take values from 1 to a maximum possible value $k_{max} = N - 1$ (self-coupling and multiple connections being prohibited). Z coincides with the Riemann zeta function in the limit $k_{max} = \inf$ and is used as normalization factor; the parameter γ is the scale-free exponent that characterizes the distribution. Variating γ is possible to variate the pendency of the distributions, favoring the presence/absence of hubs. In this analysis we compare NRBNs built with a random topology of the connections (type A) with NRBNs with scale free topology (type B). The parameters of the simulations can be found in Table 1.

The dynamical regimes: ordered vs. critical vs. disordered. As specified in the introduction, it was analytically proven that by varying some key structural parameters of RBNs it is possible to characterize the emerging dynamical regime (on average). In particular, analytical studies presented in [2] on classical RBNs (with random topology) defined a phase diagram linking two structural parameters, the average connectivity and the so-called *bias*, a parameter that defines the likelihood of having a "1" output in the Boolean functions. In particular, a parameter λ, called sensitivity [59] or Derrida exponent, allows to discriminate the different dynamical regimes. If the Boolean functions are generated randomly with a bias p and A is the average connectivity of the network we have:

$$\lambda = 2Ap(1 - p), \tag{4}$$

Values of λ smaller than 1 are typical of ordered networks, whereas values larger than 1 are characteristic of chaotic regime and the critical line at $\lambda = 1$ separate the two phases, indicating the critical region. The development of Eq. 4 for the critical case allows to draw a phase diagram regarding the dynamical regimes. In detail, for each value of p there exist a critical value of the average connectivity A such that the network is in the ordered regime for $A < A_{kr}(p)$ and in the chaotic regime for $A > A_{kr}$ The equation is the following:

$$A_{kr} = [2p(1 - p)]^{-1} \tag{5}$$

Extending the theory for RBNs with scale-free topology, as in [32], the critical γ_{kr} for a network with $p = 0.5$ can be determined according to the following equation:

$$\frac{Z(\gamma_{kr} - 1)}{Z(\gamma_{kr})} = 2 \tag{6}$$

Therefore we chose three parameters settings to design: ordered (type 1), critical (type 2), disordered (type 3) NRBNs, with both the topologies. The parameters used in the simulations can be found in Table 1.

We here present the results of the analysis on 6 classes of NRBNs: random topology-ordered regime (type $A1$), random topology-critical regime(type $A2$), random topology-disordered regime (type $A3$), scale-free topology-ordered regime(type $B1$), scale-free topology-critical regime(type $B2$) and scale-free topology-disordered regime(type $B3$).

Tree matching. After generating random NRBNs satisfying the above mentioned parameters we compare the outcoming differentiation tree with the two real differentiation trees concerning intestinal crypts and hematopoietic cells in Fig. 1.

We associate totipotent stem cells with TESs at threshold 0, cells in a pluripotent or multi-potent state (i.e. transit amplifying stage or intermediate state) with TESs with a larger threshold composed by one or more attractors, while completely differentiate cells to TESs with the highest threshold, usually composed by single attractors.

Given that each ATM can be characterizer by a high number of possible thresholds (i.e. all the distinct entries in the matrix), in order to reduce to computational costs, we do not check all the possible combination of thresholds in determining the outcoming tree. In particular, given an input tree of Y levels, we define all the possible sets of Y thresholds $\Delta_x = [\delta_1, \delta_2, ..., \delta_Y], \delta_1 < \delta_2 < ... < \delta_Y, \Delta_x \subseteq [0, 0.01, 0.02, ..., 0.15]$. Given a certain ATM, every set of thresholds Δ_x give rise to a specific tree, which is then compared with the target tree. We chose 0.15 as the maximum threshold value, because we observed that, on average, no further splitting of TESs is observed above that value (see the analyses below).

A measure of distance among the tree emerging from the NRBN dynamics and the real tree is defined as a sum of *histogram distances* for all the levels of the input tree [7][6]. The larger this quantity is the more dissimilar the input and the emergent trees are. On the contrary when this quantity is zero it is not assured that the two trees completely match, but still it is a good quantitative approximation. Considered that, given a certain NRBN, many distinct trees are possible in correspondence of different sets of thresholds Δ_x, we compare the target tree with the emerging tree showing the *minimum* tree distance.

Notice that since the set of constraints we are imposing is non-trivial, we do not expect to find many "suitable" NRBNs.

---

[6] For each level of the input tree, the algorithm compares the distribution of the number of children of the two trees. The histogram distance is then defined as the sum of the absolute value of the difference between the number of nodes in the first tree with i children in the two trees, from $i = 1$ to the maximum number of children. The overall distance is the sum of all the histogram distances of the distinct levels.

Table 1. Parameters of the Noisy Random Boolean Networks.

Common parameters				
Number of NRNBs per type		1000		
Number of nodes, N		100		
Number of initial condition per net		10000		
Maximum simulation time		10000		

Type	Description	A	p	Topology	γ
A1	Random-ordered	1.7	0.5	R	x
A2	Random-critical	2	0.5	R	x
A3	Random-disordered	2.3	0.5	R	x
B1	Scale-free-ordered	≈ 1.7	0.5	SF	2.57
B2	Scale-free-critical	≈ 2	0.5	SF	2.38
B3	Scale-free-disordered	≈ 2.3	0.5	SF	2.24

4   Results

The first analyses regard the dynamical properties of the NRBNs in the 6 cases, with specific regard to the landscape and the properties of the TESs. Notice that in all the figures the *left* panel regard the random topology networks, the *right* panel the scale-free networks, the *blue* lines concern ordered networks, the *red* lines the critical networks and the *green* lines the disordered networks.

In Fig. 3 one can see the distribution of the number of distinct attractors in the 6 cases, the left panels corresponding to the random topology case (A1, A2, A3), the right panels to the scale-free case (B1, B2, B3)[7].

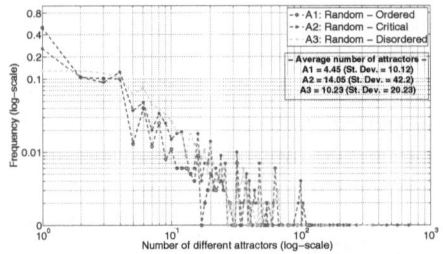 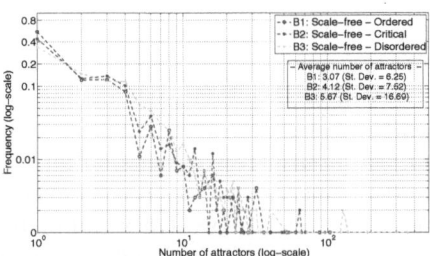

Fig. 3. Distribution of the number of distinct attractors. The distribution of the number of distinct attractors with respect to the 6 types of simulated networks is shown. The plot is in log-log scale and a box with the average value and the standard deviation of the number of distinct attractors is provided.

As expected, in NRBNs with random topology ordered networks are the most likely to have only one or a few attractors, disordered networks are the least likely to be in such situation, whereas critical networks are in between

---

[7] It is worth noticing that an attractor has always been reached within the simulation time.

the two. Conversely, despite a very high dispersion, it interesting to notice that the average number of attractors is higher for critical networks, which are however known to display a very broad range of dynamical behaviors [5]. Scale-free NRBNs show a similar overall trend with regard to the proportion of networks with a low number of attractors, yet with smaller magnitudes and, above all, with smoother differences among the three cases, hinting at slighter dynamical differences among the regimes. Furthermore, the overall number of attractors is lower in the scale-free case than in the random case, which suggest that scale-free NRBNs with analogous parameters to random topology NRBNs are indeed more ordered. This result confirms those shown in [35].

The second analysis is aimed at disentangling the effect of topology and dynamical regime on the robustness of the attractors in presence of noise. In particular in Fig. 4 one can observe the distribution (and the average) of the values of the diagonal of the ATMs in the different cases, which provides an indication on how many times a NRBN, after a single-flip perturbation, returns to the same attractor.

It can be noticed that this indicator, with respect to both the topologies, reflects the level of robustness expected from the three dynamical regimes, with ordered networks being the most insensitive to perturbations (with typically higher probabilities to remain within the same attractor), disordered networks being instead the most sensitive to perturbations (with typically low probabilities to remain within the same attractors) and critical network showing intermediate behaviors. For what concerns the case of scale-free topology, although the same ranking holds, the behaviour of disordered and critical networks is closer to the behaviour of ordered networks with either random or scale-free topology. This outcome provides another evidence of the more ordered dynamical behaviour of scale-free networks when compared to random networks.

Less predictable is the likelihood of switching toward another given attractor, which is captured by non-null off-diagonal values. It is worth stressing that this indicator differs substantially from the probability to switch toward another attractor in general, which corresponds to the negative of the probability to remain in the same attractor (and include null off-diagonal values)). The frequency of these values for the 6 cases under study are plotted in Fig. 5. Interestingly, as long as the random topology is concerned, critical networks slightly move away from both the ordered and disordered regime (whose distributions mainly overlap) and show significantly low levels of probability to switch to a given attractor. This effect cannot be observed in the case of scale-free topologies where the differences between the three distributions are nearly negligible. This result may have important effects on the emerging lineage commitment tree. It is reminded that the differentiation is modeled as a gradual improvement of the noise-control mechanisms that hinder the transitions from undifferentiated to fully differentiated cells. It might be therefore the case that, even if the probability to exit from the current attractor (which corresponds to the probability to move to any other attractor) might be considerable, if the probability to reach another given attractor is typically low, this switch is more subject to be

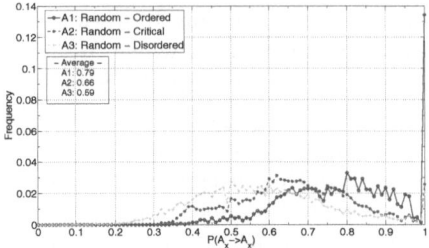 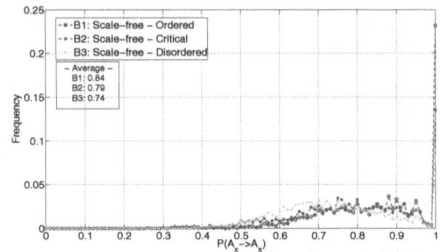

Fig. 4. Attractor stability (I). The distribution of the probability of remaining in the same attractor after a one-flip perturbation is shown, i.e. the distribution of the values of the diagonal of the ATN (bin = 0.01), w.r.t. the 6 classes of networks. In the box the average value is also shown.

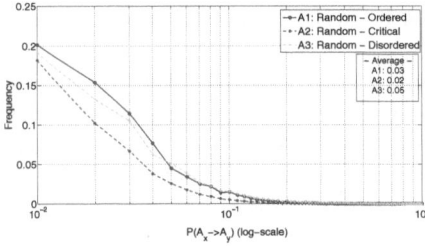 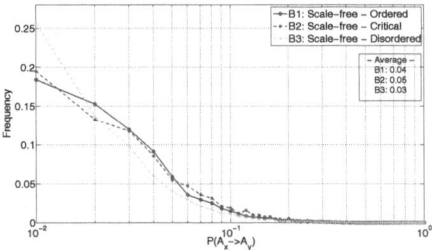

Fig. 5. Attractor stability (II). The distribution of the probability of switching from a random attractor toward another random attractor, i.e. the distribution of the non-zero off-diagonal values of the ATN (bin = 0.01) is shown with regard to the 6 typologies of networks. The x-axis is in log-scale, whereas in the box the average value is provided.

prevented by the noise control mechanism (which prevents switches that have probability below a threshold δ).

This hypothesis can be investigated by analyzing the number of TESs, which in our model corresponds to different cell types, as a function of the threshold δ. It can be indeed observed in Fig. 6 that the number of TESs of critical random network shows a steep increase in correspondence of a low δ, moving away from both random ordered and random disordered networks, which exhibit a smoother increase and finally stabilize at significantly lower values. Notice also that the higher dispersion is observed in correspondence of critical network, which suggest a larger variability in the possible behaviors and, accordingly, of differentiation hierarchies. Consistently with the results regarding the probability to switch to a given attractor (Fig. 5), the differences in the three regimes are less clear in the case of scale-free topology, where the number of TESs is typically lower. Interestingly, in the latter case, the networks showing the maximum number of TES seem to be the disordered ones, which however show an average number of TES that is not even half of those of critical random networks.

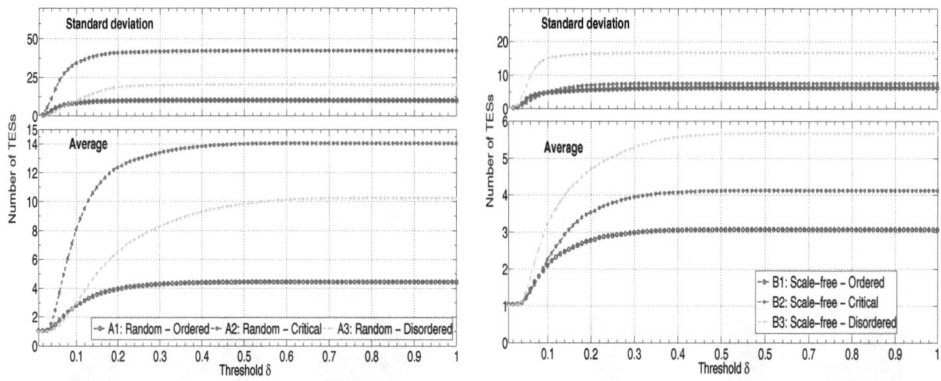

Fig. 6. Variation of the number of TESs with respect to different thresholds.
The average number (and the standard deviation) of the number of TESs with respect
to the variation of the threshold δ in the range: $\delta \in [0 : 0.01 : 1]$ is shown, with regard
to the 6 classes of networks.

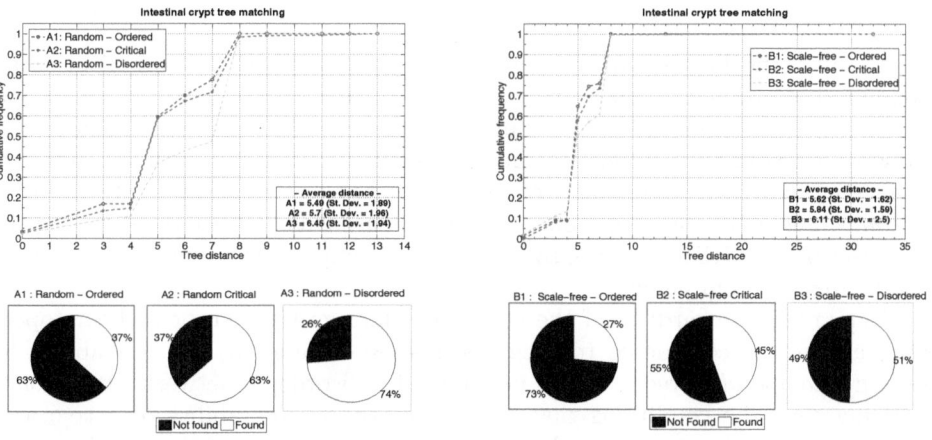

Fig. 7. Matching with intestinal crypt tree. In the upper panels the cumulative
distribution of the *minimum* tree distances resulting from the comparison between the
input differentiation tree of intestinal crypts (Fig. 1) and the suitable trees emerging
from the NRBN dynamics is provided, with respect to the different classes of networks.
Only a certain number of thresholds δ in the range: $\delta \in [0 : 0.01 : 1]$ is considered
to compute the possible trees and for each network only the minimum distance on
all the possible trees is considered. In the boxes the average values (and the standard
deviation) of the distances are shown as well. In the lower panels the percentage of
NRBNs that originate non-suitable trees for the comparison is shown such as, e.g.,
forests or trees with a number of nodes lower than the levels of the input trees.

Hence, it is possible to expect critical random network to have typically a
considerable number of different cell types (TESs) at a given differentiation level
(corresponding to a given threshold). It is therefore meaningful to investigate

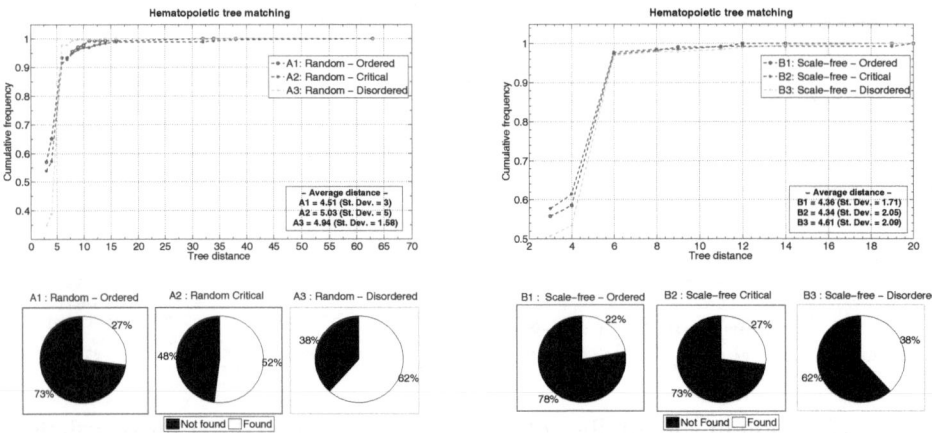

Fig. 8. Matching with hematopoietic cells tree. The comparison is made against the input hematopoietic cells differentiation tree (Fig. 1). See the previous caption for the description of the Figure.

how similar the lineage tree emerging from these networks is to real ones, as compared to ordered or disordered networks. In this regard, in Figs. 7 and 8 one can observe the cumulative distribution of the tree distance, as defined in the previous section, with respect to the two lineage trees of Fig. 1, in the different cases. Because the tree distance can be calculated only when the obtained tree is *comparable* with the target one, that is, when it has at least the same depth, the fraction of NRBNs leading to at least one comparable tree is also shown in the figures (bottom panels).

Remarkably, regardless of the considered lineage tree (crypt or hematopoietic) and of the considered topology (scale-free or random) the probability to obtain a *comparable* tree is always maximum for disordered networks. It is worth noticing that the conditions that may affect the comparability of a tree are: (i) the number of attractors, which cannot be lower of the target tree's depth; (ii) the possible existence of more than one TES at the level 0 (threshold 0) which would lead to a forest. According to the distribution probabilities in Fig. 3, disordered networks have the lowest probability to have a unique or a few attractor and therefore the lowest probability to have a tree with a smaller number of levels than the target trees.

The interpretation of the comparison of the distribution of the tree distances obtained in the comparable cases appears less straightforward. Let us first take into account the intestinal crypts differentiation tree (Fig. 7): it can be observed that the average distance does not highlight sharp differences between different dynamical regimes or topologies, and even when comparing the distribution it is difficult to tell which networks are actually performing better. Along similar lines, when the target is the hemopoietic cells tree, the differences in the six cases under study are nearly negligible. A non negligible difference can be observed only in the magnitude of standard deviations, which appears higher in random critical networks than in the other cases. The distribution of the distances regarding random critical nets shows indeed a long right tail, indicating that, in some cases, the tree emerging from these networks, despite having the same number of levels of the hemopoietic cells tree can differ substantially within each level. High distance values might be plausibly due to the typical higher number of TES observed in critical random nets.

It can however be noticed that the average distance in the intestinal crypt case (Fig. 7) is typically inferior to that of hematopoietic cells and that there is a non null probability to observe distance zero, which approximates a perfect match, in all the dynamical regimes and topologies considered. Although a distance equal to 0 is never observed in the case of hematopoietic cells tree (Fig. 8), it is however worth noticing that the a considerable fraction (the majority in case of scale-free topology) of the simulated networks can have a tree within a distance lower than 4, suggesting that the differentiation tree obtained from the sampled random networks is not that far from biologically plausible ones.

5   Conclusion

We have investigated the influence of the topology and of the dynamical regime of gene regulatory networks, as modeled with Noisy Random Boolean Network, with specific regard to the properties concerning cell differentiation. In particular, we have studied the properties of scale-free and random topology networks and in both cases we have analyzed ordered, critical and disordered networks. Remarkable differences were highlighted in the distinct cases.

First, the dynamical regime is strongly related to the stability of the attractors, suggesting that gene activation patterns in relatively more ordered networks are indeed more robust against random genomic mutations. Besides, scale-free networks appear to be generally more ordered, and hence robust, than analogous random-topology networks, so that even structurally disordered scale-free networks show behaviors that are comparable to ordered (or slightly subcritical) random networks.

A very interesting result is given by the average number and the dispersion of the number of TESs, which in our model represent cell types or modes. Random-topology critical networks show the larger average number and the highest variability, hinting at an intrinsic capability of critical network to show a great heterogeneity in the possible behaviors and, consequently, of possible emerging differentiation hierarchies. The hypothesis of criticality of gene networks was suggested by many and this could provide a further element in this direction. Conversely, and in accordance with the conclusions above, scale-free nets present a much ordered general behaviour, displaying a quite narrow range of possible behaviors, as given by the number and the variability of TESs.

The comparison with two real differentiation trees, i.e. intestinal crypts and hematopoietic cells, did not allow to achieve definitive conclusion on which network type is more suitable, even though slightly disordered networks display a larger number of comparable emerging trees, with respect to both the topologies. All in all, a surprisingly high portion of networks, across different dynamical regimes and topologies, originate trees that well approximate real ones, suggesting that rather simple differentiation schemes such those used in this study can be matched even through a random (not evolution-driven) generation. It is worth stressing that, as it has been previously observed (submitted work), networks in the slightly ordered regime can indeed exhibit behaviors that are more typical of the critical or disordered regime. Along similar lines critical or disordered dynamics can be observed in the slightly ordered regime. Hence, it could be interesting to investigate wether there exist some specific features shared by networks in the different regimes that can lead to the emergence of biologically plausible trees. Furthermore, the analysis of ensembles of networks *evolved* to fit the target trees (e.g. via bio-inspired algorithms) may unravel interesting new general properties.

Acknowledgements. This work was partially supported by the project SysBionet (12-4-5148000-15; Imp. 611/12; CUP: H41J12000060001; U.A. 53).

References

1. Alberts, B., Johnson, A., Lewis, J., Raff, M., Roberts, K., Walter, P.: Molecular Biology of the Cell, 5th edn. Garland Science, New York (2007)
2. Aldana, M., Coppersmith, S., Kadanoff, L.: Boolean dynamics with random couplings. In: Kaplan, E., Marsden, J., Sreenivasan, K.R. (eds.) Perspectives and Problems in Nonlinear Science: Springer Applied Mathematical Sciences Series. Springer, New York (2003)
3. Barabasi, A.-L., Albert, R.: Emergence of scaling in random networks. Science **286**, 509–512 (1999)
4. Blake, W.J., Mads, K., Cantor, C.R., Collins, J.J.: Noise in eukaryotic gene expression. Nature **422**, 633–637 (2003)
5. Damiani, C., Graudenzi, A., Serra, R., Villani, M., Colacci, A., Kauffman, S.A.: On the fate of perturbations in critical random boolean networks. In: Proceedings of the European Conference on Complex Systems, ECCS 2009 (Cd-rom) (2009)
6. De Matteis, G., Graudenzi, A., Antoniotti, M.: A review of spatial computational models for multi-cellular systems, with regard to intestinal crypts and colorectal cancer development. J. Math. Biol. **66**(7), 1409–1462 (2013)
7. Degasperi, A., Gilmore, S.: Sensitivity analysis of stochastic models of bistable biochemical reactions. In: Bernardo, M., Degano, P., Zavattaro, G. (eds.) SFM 2008. LNCS, vol. 5016, pp. 1–20. Springer, Heidelberg (2008)
8. Eldar, A., Elowitz, M.B.: Functional roles for noise in genetic circuits. Nature **467**, 167–173 (2010)
9. Erdos, P., Rényi, A.: On the evolution of random graphs. Bull. Inst. Int. Statist **38**(4), 343–347 (1961)
10. Furusawa, C., Kaneko, K.: Chaotic expression dynamics implies pluripotency: When theory and experiment meet. Biol. Direct. **4**, 17 (2009)
11. Hayashi, K., Lopes, S.M., Surani, M.A.: Dynamic equilibrium and heterogeneity of mouse pluripotent stem cells with distinct functional and epigenetic states. Cell Stem Cell **3**, 391–440 (2008)

12. Hoffman, M., Chang, H.H., Huang, S., Ingber, D.E., Loeffler, M., Galle, J.: Noise driven stem cell and progenitor population dynamics. PLoS ONE **3**, e2922 (2008)
13. Hu, M., Krause, D., Greaves, M., Sharkis, S., Dexter, M., Heyworth, C., Enver, T.: Multilineage gene expression precedes commitment in the hemopoietic system. Genes Dev. **11**, 774–785 (1997)
14. Huang, S., Ernberg, I., Kauffman, S.A.: Cancer attractors: A systems view of tumors from a gene network dynamics and developmental perspective. Semin Cell Dev Biol. **20**(7), 869–876 (2009)
15. Huang, S., Guo, Y.P., Enver, T.: Bifurcation dynamics in lineage-commitment in bipotent progenitor cells. Dev. Biol. **305**, 695–713 (2007)
16. Kalmar, T., Lim, C., Hayward, P., Munñoz-Descalzo Arias, S., Nichols, J., Garcia-Ojalvo, J., Martinez, A.: Regulated fluctuations in nanog expression mediate cell fate decisions in embryonic stem cells. PLoS Biol. **7**, e1000149 (2009)
17. Sundaram, S., Sundararajan, N., Savitha, R.: Introduction. In: Sundaram, S., Sundararajan, N., Savitha, R. (eds.) Supervised Learning with Complex-valued Neural Networks. SCI, vol. 421, pp. 1–30. Springer, Heidelberg (2013)
18. Kashiwagi, A., Urabe, I., Kaneko, K., Yomo, T.: Adaptive response of a gene network to environmental changes by fitness-induced attractor selection. PLoS ONE **1**, e49 (2006)
19. Kauffman, S.A.: Homeostasis and differentiation in random genetic control networks. Nature **224**, 177 (1969)
20. Kauffman, S.A.: Metabolic stability and epigenesis in randomly constructed genetic nets. J. Theor. Biol. **22**, 437–467 (1969)
21. Kauffman, S.A.: At Home in the Universe. Oxford University Press, New York (1995)
22. Kauffman, S.A., Peterson, C., Samuelsson, B., Troein, C.: Random boolean network models and the yeast transcriptional network. Proc. Natl Acad. Sci. USA **100**, 14796–14799 (2003)
23. Langton, C.G.: Life at the edge of chaos. In: Langton, C.G., Taylor, C., Farmer, J.D., Rasmussen, S. (eds.) Artificial Life II, pp. 41–91. Addison-Wesley, Reading (1992)
24. Lestas, I., Paulsson, J., Vinnicombe, G.: Noise in gene regulatory networks. IEEE Trans. Autom. Control **53**, 189–200 (2008)
25. Mc Adams, H.H., Arkin, A.: Stochastic mechanisms in gene expression. Proc. Natl. Acad. Sci. USA **94**, 814–819 (1997)
26. Miyamoto, T., Iwasaki, H., Reizis, B., Ye, M., Graf, T., Weissman, I.L., Akashi, K.: Myeloid or lymphoid promiscuity as a critical step in hematopoietic lineage commitment. Dev. Cell. **3**, 137–147 (2002)
27. Peixoto, T.P., Drossel, B.: Noise in random boolean networks. Phys. Rev. E **79**, 036108–17 (2009)
28. Raj, A., van Oudenaarden, A.: Nature, nurture, or chance: Stochastic gene expression and its consequences. Cell **135**, 216–226 (2008)
29. Ramo, P., Kesseli, Y., Yli-Harja, O.: Perturbation avalanches and criticality in gene regulatory networks. J. Theor. Biol. **242**, 164–170 (2006)
30. Ribeiro, A.S., Kauffman, S.A.: Noisy attractors and ergodic sets in models of gene regulatory networks. J. Theor. Biol. **247**, 743–755 (2007)
31. Serra, R., Villani, M., Barbieri, A., Kauffman, S.A., Colacci, A.: On the dynamics of random boolean networks subject to noise: Attractors, ergodic sets and cell types. J. Theor. Biol. **265**, 185–193 (2010)

32. Serra, R., Villani, M., Graudenzi, A., Colacci, A., Kauffman, S.A.: The simulation of gene knock-out in scale-free random boolean models of genetic networks. Netw. Heterogen. Med. **3**(2), 333–343 (2008)
33. Serra, R., Villani, M., Graudenzi, A., Kauffman, S.A.: Why a simple model of genetic regulatory networks describes the distribution of avalanches in gene expression data. J. Theor. Biol. **249**, 449–460 (2007)
34. Serra, R., Villani, M., Semeria, A.: Genetic network models and statistical properties of gene expression data in knock-out experiments. J. Theor. Biol. **227**, 149–157 (2004)
35. Serra, R., Villani, M., Agostini, L.: On the dynamics of random boolean networks with scale-free outgoing connections. Physica A: Statistical Mechanics and its Applications **339**, 665–673 (2004)
36. Shmulevich, I., Kauffman, S.A., Aldana, M.: Eukaryotic cells are dynamically ordered or critical but not chaotic. Proc. Natl. Acad. Sci. USA **102**, 13439–13444 (2005)
37. Swains, P.S., Elowitz, M.B., Siggia, E.D.: Intrinsic and extrinsic contributions to stochasticity in gene expression. Proc. Natl. Acad. Sci. USA **99**, 12795–12800 (2002)
38. Vijesh, N., Chakrabarti, S.K., Sreekumar, J.: Modeling of gene regulatory networks: A review. J. Biomed. Sci. Eng. **6**, 223–231 (2013)
39. Villani, M., Barbieri, A., Serra, R.: A dynamical model of genetic networks for cell differentiation. PLoS ONE **6**(3), e17703 (2011). doi:10.1371/journal.pone.0017703
40. Warren, L., Bryder, D., Weissman, I.L., Quake, S.R.: Transcription factor profiling in individual hematopoietic progenitors by digital RT-PCR. PNAS **103**(47), 17807–17812 (2006)
41. Yamanaka, H.: Elite and stochastic models for induced pluripotent stem cell generation. Nature **460**, 49–52 (2009)

Molecular Communication Technology: General Considerations on the Use of Synthetic Cells and Some Hints from In Silico Modelling

Fabio Mavelli[1], Giordano Rampioni[2], Luisa Damiano[3], Marco Messina[2], Livia Leoni[2], and Pasquale Stano[2(✉)]

[1] Chemistry Department, University of Bari, Bari, Italy
[2] Science Department, Roma Tre University, Rome, Italy
pasquale.stano@uniroma3.it
[3] Centre for Research on Complex Systems, University of Bergamo, Bergamo, Italy

Abstract. Recent advancements in synthetic biology pave the way to the design and construction of synthetic cells of increasing complexity, capable of performing specific functions in programmable manner. One of the most exciting goal is the development of a molecular communication technology based on the exchange of chemical signals between synthetic and natural cells. We are currently involved in such a research program. Following our previous contributions to WIVACE workshops (2012–2013), here we present the project, and discuss some general considerations on the use of synthetic cells for developing novel bio-chemical Information and Communication Technologies (bio-chem-ICTs). Moreover, by analysing in detail a mathematical model of synthetic cell/natural cell communication process, we provide some hints that can be valuable for the next experimental steps.

Keywords: Bio-chem-ICTs · Molecular communication · Quorum sensing · Synthetic biology · Synthetic cells

1 Introduction

Synthetic biology (SB) can be defined as (a) the design and construction of new biological parts, devices, and systems, and (b) the re-design of existing, natural biological systems for useful purposes (http://syntheticbiology.org). Classical SB approaches, therefore, generally consist in the integration of biology and engineering [1,2], and aim at designing biological organisms (generally microorganisms) by the addition, elimination, modification, and redesign of parts and circuits. However, a novel SB paradigm is emerging, and it foresees in the construction of synthetic cells from their components (Fig. 1a). Such an approach is often referred as "bottom-up", and consists in assembling synthetic cells (from molecular components like DNA, proteins, lipids, etc.) endowed with a minimal degree of complexity, yet capable of displaying cell-like functions - and ultimately of being alive [3].

© Springer International Publishing Switzerland 2014
C. Pizzuti and G. Spezzano (Eds.): WIVACE 2014, CCIS 445, pp. 169–189, 2014.
DOI: 10.1007/978-3-319-12745-3_14

a)   b)

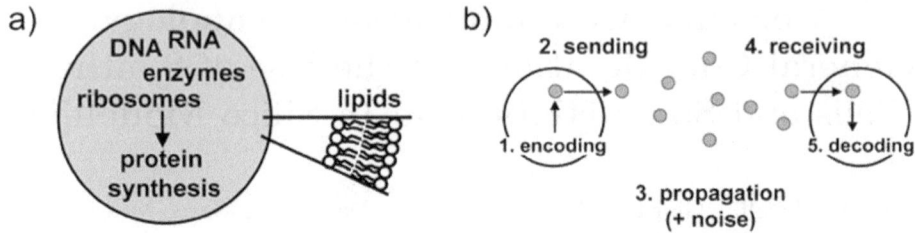

Fig. 1. (a) Synthetic cells [3] can be built by encapsulating the minimal number of biomolecules inside lipid vesicles (liposomes). To date, synthetic cells can synthesize functional proteins with good efficiency, but are unable to generate energy, grow/duplicate. (b) Outline of a molecular communication process based on the diffusion of signal molecules from a sender cell and a receiver cell, sharing the general scheme of a Shannon communication model [7]. Reproduced from [6] with the permission of Springer.

The construction of synthetic cells is rapidly progressing owing to the recent convergence of liposome technology and cell-free transcription-translation (TX-TL) systems. In particular, synthetic cells are obtained by encapsulating TX-TL systems inside lipid vesicles (liposomes), together with the gene(s) of interest. The encapsulated TX-TL system produces the proteins of interest, which in turn perform the desired function within the synthetic cells. When built in this way, synthetic cells become a very interesting tool for investigating scientific open questions (for example, the origin of life) and for developing novel biotechnological tools. Their design is versatile and modular, they are bio-compatible, and, importantly, fully programmable (at least in principle). By constructing well-designed synthetic cells it becomes possible to observe, study and control the dynamics of biochemical, sensorial or regulatory networks without the interference of background processes that are always present in biological cells.

Molecular communication (Fig. 1b), being a fundamental and widespread process in biological systems, is a fascinating target function for synthetic cells [4–6], that has not been realized yet. By processing molecular signals, synthetic cells could effectively communicate with each other and, intriguingly, with biological cells. This would pave the way to the construction of novel microscopic systems based on embodied self-production and self-maintenance (autopoiesis), and operating in the domain of bio-chemical Information and Communication Technologies (bio-chem-ICTs) [7–9]. Such a project, in our opinion, represents a starting point not only for future advancements in nanomedicine [10], but also for theoretical and experimental developments concerning the concept of minimal (embodied) autopoietic cognition, which would open novel perspectives in the synthetic study of natural cognition and in artificial intelligence [11]. Synthetic cells can be considered as molecular robots that can be functionalized with sensors and actuators. In contrast with electromechanical robots, however, synthetic cells functions are embodied in the structure of their molecular components, which continuously and parallelly interact with each other on the basis of diffusion, molecular recognition, and cooperative behaviour – without the need of a central directional unit.

In previous contributions we introduced the central idea of endowing synthetic cells with minimal communication devices, and proposed a blueprint for the experimental construction of such systems. Following the usual synthetic biology work-flow (design, modelling, construction), we have also recently presented an in silico model of unidirectional synthetic-to-natural cell communication [6]. Here we would like to further promote our vision, presenting the synthetic communication project. Firstly, we will shortly introduce a synthetic biology approach for developing synthetic cells, and review it according to the general concepts in molecular communication technologies, which have been recently illustrated by T. Nakano, A. W. Eckford, and T. Haraguchi in their monography [9]. Then, we will present a mathematical model that describes synthetic cell/natural cell communication [6], supplied with a detailed analysis of its parameters.

2   The Experimental Approach

In the past two years we have been working on the extension of the current synthetic cell approaches toward the design of molecular communication systems based on the synthesis and the exchange of diffusible chemicals. The first experimental approach is due to Ben Davis and collaborators [12], who reported a simple chemical communication system based on the encapsulation of the so-called formose reaction inside liposomes. As a result of internal reaction, a chemical was produced by the synthetic cell. Once released in the medium, this chemical compound can reach bacteria and activate a biological response (chemoluminescence). Davis' synthetic cells, however, lack modularity, programmability and variability in their design, because they are based specifically on a certain particular chemical reaction (the formose reaction). A synthetic biology approach appears much more powerful. In fact, since it is based on protein synthesis inside liposomes [13], almost all types of synthetic cells can be envisaged, and their construction made possible by employing biological elements like enzymes, receptors, transcription factors, etc. These molecular components are endowed with a natural chemical processing capability and can be combined modularly so to construct multifunctional devices.

2.1   Synthetic Cells

Synthetic cells (or, more precisely, semi-synthetic minimal cells [3]), are cell-like compartments based on biochemical machineries encapsulated inside lipid vesicles (liposomes) (Fig. 1a) [14]. In order to build synthetic cells capable of non-trivial behaviour, several macromolecules (DNA, RNA, proteins) must be inserted inside liposomes. This can be conveniently done by employing a rather novel preparation method called *droplet transfer* method [15–18], which allows an efficient encapsulation of solutes inside giant vesicles (GVs). In turn, GVs are conveniently manipulated and visualized thanks to their very large size (diameter $>1\,\mu m$). As TX-TL machinery, the PURE system [19] is often used because

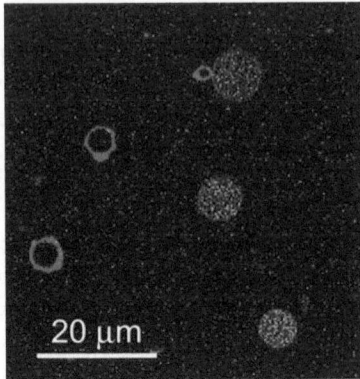

Fig. 2. PURE-system containing synthetic cells based on giant lipid vesicles are able to produce the green fluorescent protein in their inner aqueous core (green fluorescence). The liposome membrane (red fluorescence) has been stained with Trypan Blue. Reproduced from [6] with the permission of Springer (Color figure online).

it contains the minimal number of molecular components necessary for *in vitro* protein synthesis, all well characterized and present in known concentrations. This feature makes the PURE system the perfect tool for synthetic biology, *i.e.*, according to the concept of *standard parts* (http://partsregistry.org). Several proteins and enzymes have been synthesized inside lipid vesicles of various type by using the PURE system [14], including membrane enzymes [20], see Fig. 2. Thus, current synthetic cell technology allows the construction of cell-like compartment of non-trivial complexity that can perform specific functions. It is realistic to think that the enzymes, the receptors and other molecules that take part to encoding/sending or receiving/decoding circuitry can be synthesized inside synthetic cells by TX-TL reactions.

This scenario corresponds to the construction of synthetic cells capable of producing and manipulating chemical information, and therefore establishing chemical communication between each other or with natural biological cells. Since the machineries for performing these actions (computations) are themselves produced in – and by – the synthetic cell, possibly under control of other chemical information systems, it is evident that synthetic cells, when properly developed, will represent the prototype of autonomous artificial systems. Such an approach is therefore one of the best candidates for a novel paradigm in artificial life and artificial intelligence.

2.2 Bacterial AHLs as Molecular Communication System

In the context of developing communicating synthetic cells with minimal complexity, we believe that exploiting the bacterial communication systems is the most advantageous choice to start with. The bacterial world offers a plethora of communication systems well characterized at the molecular level, hence exploitable for synthetic biology approaches. Indeed, most bacterial genera coordinate their activities at the population level via a widespread intercellular communication system known as quorum sensing (QS) [21–24]. Briefly, QS relies on the synthesis, secretion/diffusion, reception and decodification of signal molecules by members of a bacterial community. In order to design a minimal communication mechanism inspired from cell communication protocols, a proper design in terms of choice of molecular parts and devices to be implemented in synthetic cells is needed. QS systems based on N-acyl-homoserine lactones (AHLs) as signal molecules are promising candidates to establish a synthetic/natural communication system. In particular, AHLs synthesis requires a single step reaction usually driven by a single enzyme, and, when they permeate into the cell, AHLs are perceived by a cytoplasmic receptor. Moreover, AHLs are considered to be free-diffusible across biological membranes.

3 Molecular Communication Technology

The technology of molecular communication is a novel research branch in rapid expansion. Essentially, it deals with the manipulation of chemical information in biology and chemistry, and promises interesting application, especially in nanomedicine. Among many others, Tadashi Nakano and collaborators presented a series of programmatic papers [25] on the employment of molecular communication mechanisms that intriguingly inspired us. In particular, we were surprised to see how well our synthetic cell approach [3, 14] fits with the scenario presented by other authors.

Molecular communications occur in all biological systems at the intra-cellular and inter-cellular levels. There are several examples taken from unicellular and multicellular organisms, for example the regulation of transcription, the functioning of surface receptors, the pathways of calcium signalling, the hormonal and neuronal signalling, and several others. In the context of bio-chem-ICTs, molecular communication is implemented in nano- and micro-machines (systems) – made of biological or chemical materials – that can perform computation, sensing, actuation. Shannon theory also applies to molecular communication, which requires encoding, sending, propagation, receiving and decoding steps. However, in contrary to familiar telecommunication systems, molecular communication is endowed of peculiar traits that are summarized in Table 1. It is useful to quickly comment, following the discussion presented in [9], some of the entries in Table 1, with specific reference to the project proposed here.

The first three entries are self-explanatory. The creation of a communication channel between synthetic and natural cells is necessarily based on molecules, those that can be recognized by the biological cell as meaning-carrying ones. The synthetic cell, as a whole, act as a complex micro-machine, composed by several thousands of biomolecules belonging to more than 100 different species. For example, synthetic cells like those shown in Fig. 2, with diameter around $5\,\mu m$, contain several thousands of ribosomes and TX-TL enzymes, millions of low molecular weight species. The membrane – which is itself built by million

Table 1. Comparison between traditional (electro-optical) and chemical communications, adapted from [9, 25]

		Traditional communication	Molecular communication
1	Communication carrier	Electromagnetic waves	Molecules
2	Devices	Electronic devices	Bio-nanomachines
3	Signal type	Electronic and optical	Chemical
4	Encoded information	Voice, text, video	Phenomena and chemical states
5	Behaviour of the receiver	Interpretation	Chemical reactions
6	Propagation media	Air/cables	Aqueous
7	Propagation speed	Speed of light	Extremely slow
8	Propagation range	m-km	nm-μm
9	Energy consumed	Electrical (high)	Chemical (low)
10	Other features	Accurate	Bio-compatible, parallel, stochastic

lipid molecules kept together by non-covalent interactions – encloses all these molecule together and confines them in the inner vesicle lumen. This complex assembly operates within the realm of physics and chemistry but its functions are biochemically controlled, *i.e.*, information is processed as it happens in biological cells. It can be designed so to enzymatically produce an output (a molecule) that convey a certain meaning for the biological partner. This is a chemical signal (entry nr. 3) embodied as a molecule with well-specific three-dimensional structure.

Once synthesized and released in the environment, the chemical species will diffuse around creating a three-dimensional "chemical field" which can excite the sensors of the receiving biomachine, for example a living cell. The signal molecule will communicate, to the receiver, the chemical state of the sender (*i.e.*, a state where the production of a certain *signal synthase* is ON). In more complex examples, depending on the cellular internal state, different kinds of signals, combination of signals, or signals of different amplitudes could convey the message, see Fig. 3.

Moving to entry nr. 5, the chemical signal will be able to trigger one or more chemical reactions in the receiver cell, and these reactions will constitute the response to the signal.

Low speed and short range are two typical features of molecular communications in aqueous media (entries 6–8). In fact, when the communication is based on a free diffusion mechanism, as in the case of synthetic cell sending a signal molecule to biological cell, it is expected that the distance between the sender and the receiver would play a major role. The average time τ required for a molecule to propagate over a distance x can be described by the equation $\tau \approx x^2/D$, where D is the diffusion coefficient. Therefore τ scales with the square of the distance, and doubling the distance means quadruplicating the travelling time. In turn, D depends on the size and shape of the molecule. Large molecules, having low D, will travel slower than small ones. Since in our case the signal molecule will be a quite small one (molecular weight of about 200–400 Da), using

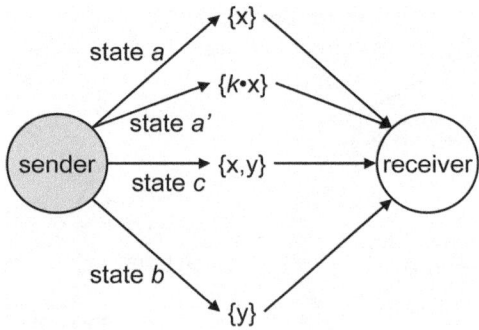

Fig. 3. Illustration of four simple ways of meaning-generation in molecular communications between synthetic/synthetic or synthetic/natural cells. State a is encoded by the signal molecule x; state a' by a different amount of signal molecule $k \cdot x$; state b by the signal molecule y; state c by a combination of x and y. Note, however, that it could be difficult to distinguish between state a and a' because the amount of signal molecule x in a certain point of space depends on the distance of the sender(s) from that point. See also [9] for additional examples (*i.e.*, signalling with timing).

an estimated diffusion coefficient D of $500\,\mu m^2/s$, it turns out that one molecule would take about $\tau \sim 20\,s$ when $x \sim 100\,\mu m$ (an estimate of the expected distance between senders and receivers in a plausible experimental setup, see below). The accumulation of signal molecules in the receiver, which is necessary to trigger its response, will depend, therefore, by the geometry of the system, and also by the rate of signal production. Considering that biological responses of the TX-TL type have characteristic response times of the order of 10^0–10^2 min it can be concluded that free diffusion in aqueous medium, despite its slowness, will probably not limit the mechanism we would like to exploit (Fig. 1b). As it will be shown below, however, the steps before and after the free diffusion in the medium can be rate-limiting. Signal molecules must be released by the transmitter and taken up by the receiver, and in both cases they have to cross the boundaries (the lipid membranes) of these systems.

From the energetic viewpoint (entry nr. 9), as all biological micro- and nanomachines, synthetic cells are powered by chemicals. In our specific case, TX-TL reactions consume nucleoside triphosphates to build the messenger RNA as well as to carry out protein synthesis. The energy requirement for TX-TL reaction, therefore, is significant. Since synthetic cells do not autonomously produce chemical energy, energy-rich compounds must be included in the encapsulated reaction mixture, and this is the usual way to proceed in current research. Alternatively, energy-rich compounds could be given externally by providing a route for entering the synthetic cell. According to previous results [26] it is sufficient to assemble an α-haemolysin pore on the vesicle membrane – Fig. 4 – to allow the entrance of low molecular weight compounds ($<3\,kDa$), such as the energy-rich compounds, without releasing macromolecules. Another possible, more complex but also more intriguing solution could be endowing synthetic cells with

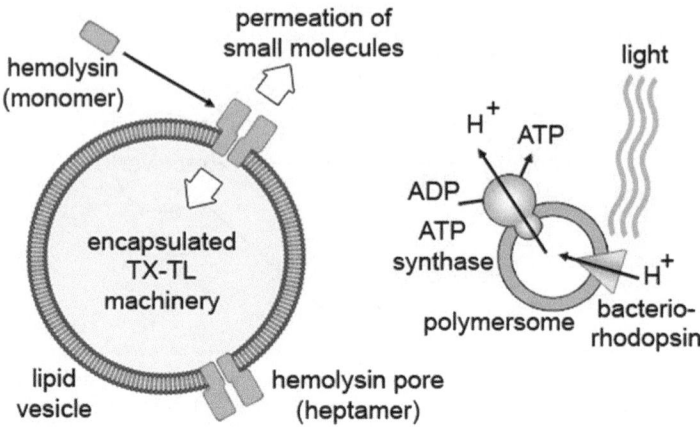

Fig. 4. (a) Formation of α-hemolysin pore allows the exchange of small molecule across the liposome membrane, as shown in [26]. (b) The encapsulation of energy-producing polymersomes [27] inside a giant vesicle could provide to the latter a sort of organelle that produce energy (ATP) after irradiation.

an energy producing system. Nano-polymersomes functionalized with bacteriorhodopsin and F_0F_1-ATP synthase have been described [27]. These particles, when inserted inside GVs containing the TX-TL mixture, could synthesize ATP after formation of a local chemiosmotic H^+ gradient, exploiting light energy as primary source (Fig. 4). In other words, such particles should function as synthetic energy-producing organelles inside the synthetic cell. Note that the assembly of multi-compartment systems with this architecture (vesosomes) is already known [28,29].

The final three features concern stochasticity, parallelism and biocompatibility (entry nr. 10). The latter property is well evident and do not require any comment (being synthetic cells composed of biomolecules).

Parallelism is obtained by letting different molecular circuitries working simultaneously, and this concept is linked to the synthetic biology issue of modularity. Synthetic cells can be designed and built in order to have modular subsystems capable of performing operations with minimal interference. The first example of modular process is given by the TX-TL reactions, which in the PURE system are implemented in four modules, namely, 1. transcription, 2. translation, 3. aminoacylation, 4. energy regeneration. Genetic circuits can be designed in order to operate in parallel [30], and their performance can be fine tuned, at least in principle, by accurate re-design of the promoter regions.

Stochasticity is an inherent characteristic of molecular nano- and microsystems [31]. In the context of synthetic/natural cell communication, stochastic effects play a role at three different levels. Firstly, stochastic effects will determine the composition of synthetic cells during their assembly (extrinsic stochastic effects). It is well known that the construction of synthetic cells rarely brings about very homogeneous populations. The compositional diversity among individual vesicles, *i.e.*, the diversity of their inner content [32], is high and generates – in turn – heterogeneity in the biochemical mechanisms of encoding, production and exportation of chemical signals, as well as signal reception and processing. In principle, this source of stochasticity can be reduced by assembling synthetic cells with methods allowing a more accurate control of synthetic cell content. The intrinsic stochasticity of molecular reaction networks is the second aspect, and applies both to synthetic and biological counterparts. This cannot be eliminated because derives from the unpredictability of molecular motion and of the reaction events, especially when systems are built with a limited number of molecules. Intrinsic stochasticity generates different reaction dynamics even within a set of identical molecular machines. Thirdly, stochastic factors affect the efficiency of molecular communication channel and of the signal-to-noise ratio. Free diffusion is a stochastic phenomenon and the time required to the signal molecule to reach the target is a random variable. The signal-to-noise ratio is also affected by stochasticity, and a robust molecular communication system is obtained by sending a high number of signal molecules (signal strengthening). The noise, on the other hand, is associated to the environmental presence of molecules with a structure similar to the signal molecule, or of signal molecules present in the environment following a previous communication event. The degradation or the removal of these unwanted molecules can therefore improve the efficiency of molecular communication (noise weakening).

4   In Silico Modelling

The use of mathematical models is a major feature of synthetic biology, because a mathematical description of the system under study helps to quantitative thinking and supports a bioengineering approach. The system under study consists in a synthetic cell sending a chemical signal to a bacterium (Fig. 5).

The model refers to a realistic case of synthetic cells producing a short chain and therefore freely permeable AHL that is recognized by bacteria as a chemical messenger. In particular, after PURE system encapsulation, vesicles should be able to produce the RhlI enzyme, which in turn will catalyse the synthesis of N-butyryl homoserine lactone (C_4-HSL) from its precursors. The biological partner is instead an engineered *Pseudomonas aeruginosa* strain. The system is composed by a number of bacteria N_{bact} and of synthetic cells N_{sc} (in a 320-to-1 fixed ratio) dispersed in a volume of 0.2 nL. Synthetic cells have a diameter of about 5.4 μm. The processes occurring inside synthetic cells and inside bacteria are described by a set of ODEs (Eqs. 1–10), shaped according to Michaelis-Menten equations [33] and diffusion rates. The thermodynamic and kinetic constants required to numerically solve the ODEs set were found in the specialized articles on AHL QS systems, in databases like BRENDA and B10NUMB3R5, or estimated as educated guesses. Further details can be found in [6].

As shown in Fig. 5 two biochemical steps takes place in the synthetic cell, namely, ① the production of a synthase E, and ② the enzymatic reaction

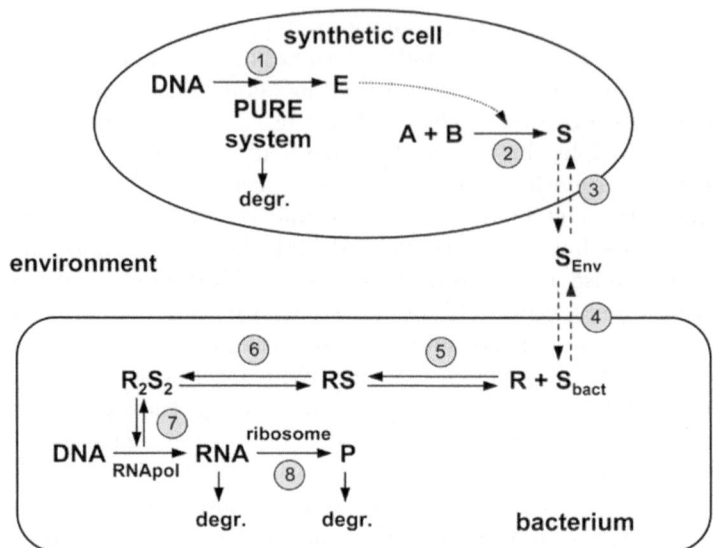

Fig. 5. Simplified model of synthetic cell sending a freely diffusible chemical signal (S) to a natural cell. Reproduced from [6] with the permission of Springer.

that produces S (the signal molecule) from the substrates A and B. These two processes are modelled by the ODEs nr. 1–2.

$$\frac{d[\text{E}]}{dt} = (k_{TXP}St)\, k_{TLP}S \cdot e^{-k_{inact}PSt} \tag{1}$$

$$\frac{d[\text{A}]}{dt} = \frac{d[\text{B}]}{dt} = -k_{cat}[\text{E}] \frac{[\text{A}]}{K_{MA} + [\text{A}]} \frac{[\text{B}]}{K_{MB} + [\text{B}]} \tag{2}$$

In particular, to model step ① we followed the experimentally derived profile obtained by the Yomo's group (details in [34]), whereas the parameters reported by [35] were used to set up the Michaelis-Menten rate equation of step ②.

Two diffusion processes, modelled by ODEs nr. 3–4 and by the first term on the right-hand side of ODE nr. 5, bring the signal molecule from the sender nanomachine (the synthetic cell) to the receiver one (the bacterium). We have modelled the diffusive steps as simple gradient-driven permeation because the S molecule we have in mind (a short chain AHL) is considered a freely diffusible species [36]. The membrane permeability $\wp$ has been set to $0.1\,\mu\text{m/s}$.

$$\frac{d[\text{S}]_{sc}}{dt} = k_{cat}[\text{E}] \frac{[\text{A}]}{K_{MA} + [\text{A}]} \frac{[\text{B}]}{K_{MB} + [\text{B}]} - \frac{\sigma_{sc}\wp}{V_{sc}} \left([\text{S}]_{sc} - [\text{S}]_{env}\right) \tag{3}$$

$$\frac{d[\text{S}]_{env}}{dt} = N_{sc} \frac{\sigma_{sc}\wp}{V_{sc}} \left([\text{S}]_{sc} - [\text{S}]_{env}\right) - N_{bact} \frac{\sigma_{bact}\wp}{V_{bact}} \left([\text{S}]_{env} - [\text{S}]_{bact}\right) \tag{4}$$

Reaching the bacterium by diffusion (process ④ in Fig. 5) the signal molecule S activates a series of steps, namely,

- ⑤ binding to the receptor R to give the non-covalent complex RS
- ⑥ non-covalent dimerization of RS to give the transcription factor R_2S_2
- ⑦ binding of R_2S_2 to the promoter region and activation of the transcription (operated by RNA polymerase)
- ⑧ translation of the messenger RNA produced in ⑦ to give the reporter protein P (operated by ribosomes)

By reporter protein we refer to a protein that can be easily detected in the bacterium, for example a fluorescent protein, or an enzyme catalyzing the formation of fluorescent or luminescent products. In this model, the reporter protein is 250 amino acid long. Moreover, two additional degradation reactions (messenger RNA and reporter protein) are considered. This reaction pattern is described by ODEs nr. 5–10

$$\frac{d[S]_{bact}}{dt} = \frac{\sigma_{bact}\wp}{V_{bact}}\left([S]_{env} - [S]_{bact}\right) - k_{on}[R][S]_{bact} + k_{off}[RS] \tag{5}$$

$$\frac{d[R]}{dt} = -k_{on}[R][S]_{bact} + k_{off}[RS] \tag{6}$$

$$\frac{d[RS]}{dt} = k_{on}[R][S]_{bact} - k_{off}[RS] - 2k_{dim}[RS]^2 + 2k_{diss}[R_2S_2] \tag{7}$$

$$\frac{d[R_2S_2]}{dt} = k_{dim}[RS]^2 - k_{diss}[R_2S_2] \tag{8}$$

$$\frac{d[RNA]}{dt} = \frac{k_{TX}C_{RNApol}}{3L}\frac{C_{DNA}}{K_{DNA} + C_{DNA}}\frac{[R_2S_2]^n}{K_{R2S2}^n + [R_2S_2]^n}$$
$$\qquad - k_{deg,RNA}[RNA] \tag{9}$$

$$\frac{d[P]}{dt} = \frac{k_{TL}C_{rib}}{L}\frac{[RNA]}{K_{RNA} + [RNA]} - k_{deg,P}[P] \tag{10}$$

A detailed discussion on how the about 40 parameters included in the model have been derived can be found in [6]. Most of them refer to *Escherichia coli*.

Simulation results are shown Fig. 6. The enzyme production inside the synthetic cell reaches a value of about $1\,\mu M$, in good agreement with the experimental findings for PURE system TX-TL reactions. Once synthesized, E catalyzes the combination of A and B to give S, the signal molecule. A peak of S inside synthetic cell (ca. $0.8\,\mu M$) appears at about 2.5 h, due to the fact that S accumulates inside synthetic cell and is slowly released in the environment, which acts as a sink, determining a concentration of S of about $0.2\,\mu M$.

The high affinity of the signal molecule for its receptor S ($K_d = 10^{-3}\,\mu M$) and the efficient dimerization ($K_{dimerization} = 10^3\,\mu M^{-1}$) bring about a ready formation of the transcription factor R_2S_2. Its binding to the promoter has been described as cooperative, with an estimated Hill coefficient of 1.5 (*cf.* Eq. 9). Following these preliminary steps, transcription and translation finally start, producing in sigmoidal fashion both messenger RNA and reporter protein. A plateau [P] value of about $0.4\,\mu M$ is reached after 4 h.

Although the outcomes of this simulation study actually encourage us to proceed with the experimental approach, these results are based on some parameters

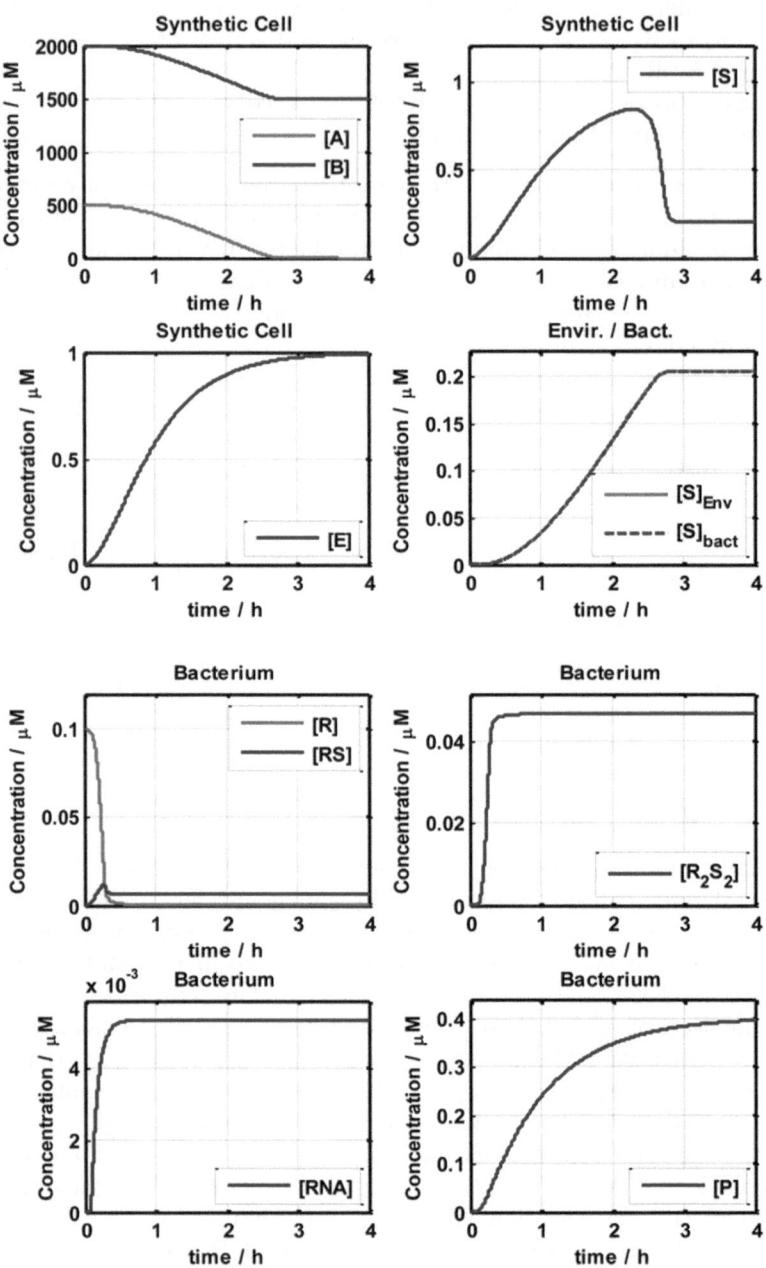

Fig. 6. Results of numerical integration (the ODEs nr. 1–10). Reproduced from [6] with the permission of Springer.

that have been estimated. In the next section, we will show what would be the result, in terms of kinetic and amount of reporter protein production, when these critical parameters are changed within a range of values. A positive outcome is considered only when a detectable amount of reporter protein is produced by the bacteria (considering a detection limit of around $0.05\,\mu M$).

4.1   Parameters Scan

In order to assay whether the conclusions obtained by the model are robust with respect to our choices of its kinetic/thermodynamic parameters, we carried out a study on how the production of the reporter protein is affected by the parameter estimated values. This was done by keeping constant all parameters found in previous section (reported in [6]) and change only one of them. In particular, we focused on the following estimated parameters: (i) the permeability $\wp$ of the signal molecule across the lipid bilayer; (ii) the initial concentration of the receptor R in the bacterium ($[R_0]$); (iii) the efficiency of RS formation (R + S $\rightleftharpoons$ RS), by changing the association rate constant (k_{on}); (iv) the dimerization efficiency (2 RS $\rightleftharpoons$ R$_2$S$_2$) by changing the dimerization rate constant (k_{dim}).

The permeability $\wp$ strongly affects the timing of bacterial response, as shown in Fig. 7. When varied from 10^{-6} to 10^{-10} cm/s several effects are clearly visible in the dynamics of the receiving system. In particular, the formation of RS, and therefore of R$_2$S$_2$, occurs at different time scales and in the case of extremely low $\wp$ (10^{-10} cm/s) the amount of R$_2$S$_2$ is limited to about 1 nM. Nevertheless, in all cases, the transcription is activated in a sigmoidal fashion, essentially because of the expected high affinity of R$_2$S$_2$ for the promoter region on the DNA ($\sim$25 pM). However, the RNA and protein syntheses are delayed more and more as the permeability coefficient decreases. Despite this, in all cases it results that bacteria produce an amount of reporter protein higher than $0.05\,\mu M$ (the estimated threshold value) within about 2 h.

In contrast to the values of $\wp$, scanning by orders of magnitude the (ii–iv) parameters does not affect significantly the production of the reporter protein, as illustrated in Figs. 8, 9 and 10. In all cases a variation of RS and R$_2$S$_2$ concentration is clearly evident, but this does not suffice to significantly change the bacterial transcription, and consequently the bacterial translation. For example, the amount of the receptor R in the bacterium could be much less than what we supposed to be ($0.1\,\mu M$). In this case, the RS and R$_2$S$_2$ concentrations will be correspondingly reduced, but such a change will affect the TX-TL reaction in negligible way. The variation of the other parameters can be described accordingly.

From these results it is evident that even order-of-magnitude variations in the estimated parameters ($\wp$, $[R]_0$, k_{on}, k_{dim}) do not significantly change the main conclusions that can be drawn by observing Fig. 6, with the exception of a significant timing effect when $\wp$ is strongly reduced. The protein synthesis is mainly controlled by the rate of transcription, which – in our model – is hardly affected by variations of R$_2$S$_2$ concentration, at least in the range of explored

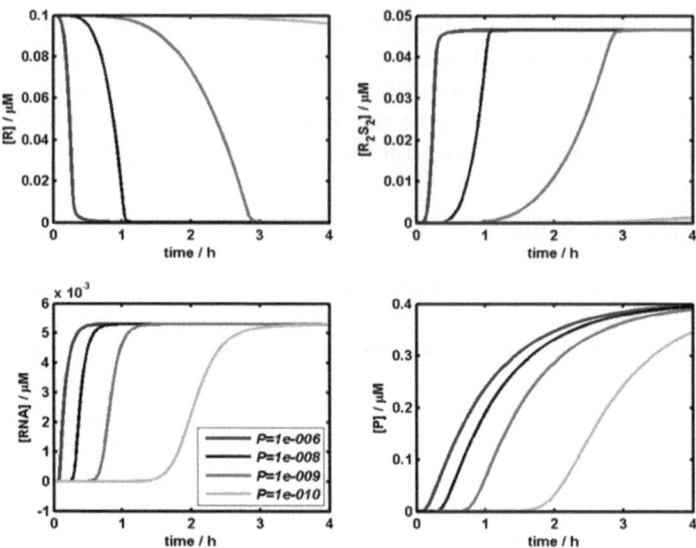

Fig. 7. Variations of bacterial response with the permeability coefficient ($\wp$, cm/s): value in the model 10^{-5}; scan from 10^{-6} to 10^{-10}.

parameters. The influence of the permeability constant is actually more related to physical reasons than to the biochemistry of the response circuitry. In order to better evidence this behaviour we plotted the rates of the processes occurring in our system against time (Fig. 11).

The rates of enzyme and signal molecule production inside synthetic cell are both high, do not change very much in time, and strongly correlate with each other. The rate of S release in the environment is somehow related to the rate of signal production in the synthetic cell, but it more than 100 times slower. It increases in time as S is synthesized, according to the generation of a larger chemical gradient across the synthetic cell membrane. On the other hand, as expected, the rate of R_2S_2 production, transcription and translation are correlated with each other, and the last two process in a very close manner (see arrows in Fig. 11). At short time, these three rates are all very low, but afterwards rapidly increases all together. Their behaviour diverges at long times because of the interplay of different reactions (mRNA and P degradation).

In conclusion, although our model is based on the estimation of several unknown kinetic and thermodynamic parameters, it results quite robust against changes of these estimates, mainly because of the estimated very high affinity of the transcription factor R_2S_2 for the DNA promoter. The permeability of the signal molecule is also a key factor determining the appearance of the reporter protein in the receiver cell (bacterium).

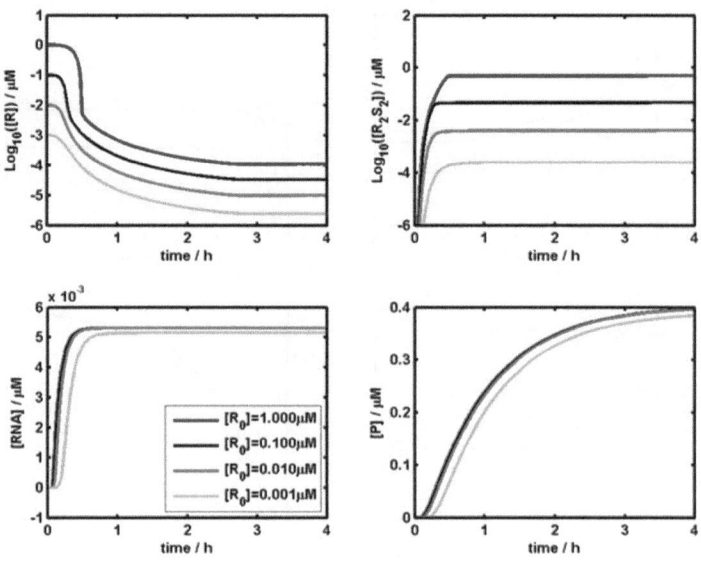

Fig. 8. Variations of bacterial response with the receptor R initial concentration ($[R_0]$, μM): value in the model 10^{-1}; scan from 10^{-3} to $10^0\,\mu$M.

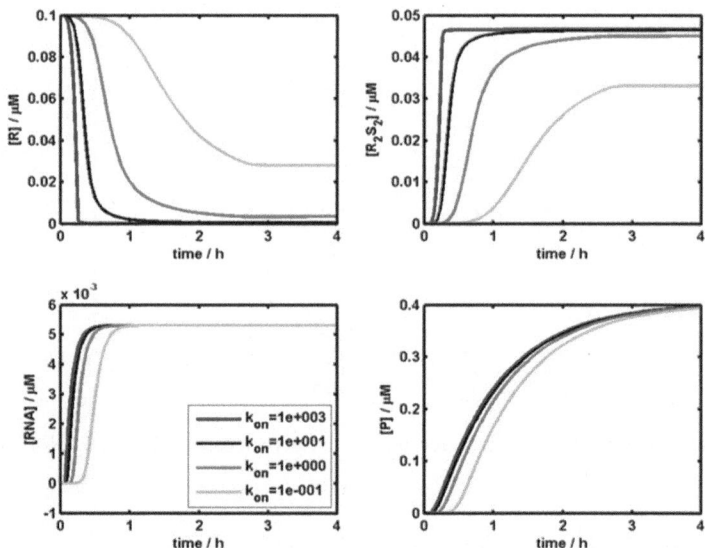

Fig. 9. Variations of bacterial response with the rate constant of RS formation (k_{on}, $\mu M^{-1}\,s^{-1}$): value in the model 10^2; scan from 10^3 to 10^{-1}.

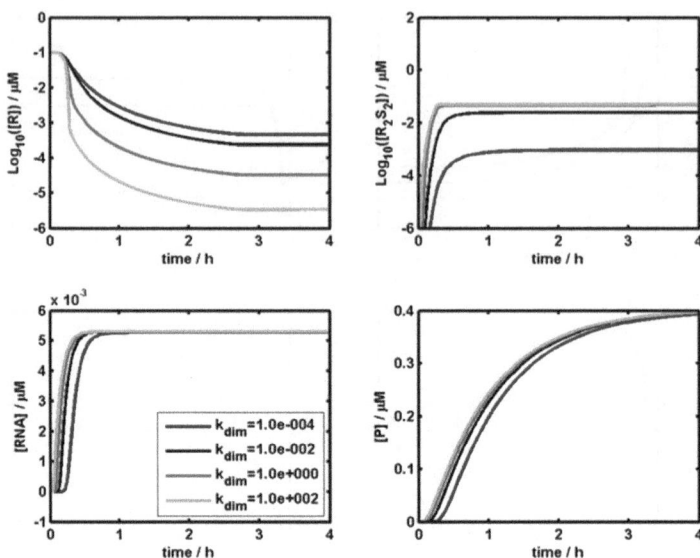

Fig. 10. Variations of bacterial response with the RS dimerization rate constant (k_{dim}, $\mu M^{-1} s^{-1}$): value in the model 10^0; scan from 10^{-4} to $10^2 \mu M^{-1} s^{-1}$

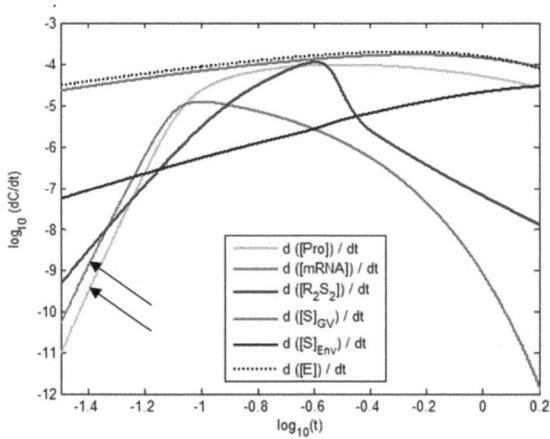

Fig. 11. Rates versus time plot referred to some processes of Fig. 5. Note the bilogarithmic scale.

5 Including the Spatial Information

Although it has not been explicitly said, the diffusion of S molecule is considered and modelled as only occurring across the membranes of synthetic cells and of bacteria, *i.e.*, the diffusion *through* the environment is considered as instantaneous. The model is built as a well-stirred reactor, meaning that – within each of the three compartments (synthetic cell, environment, bacteria) – the

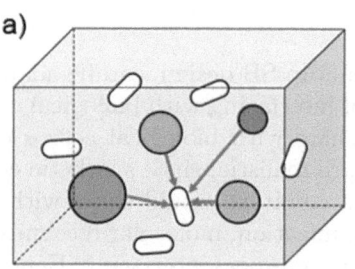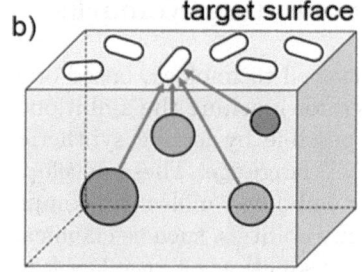

Fig. 12. Two possible way of let synthetic cells (green) and bacteria (white) interact by exchanging chemical signals, for example by a signal molecule sent by synthetic cells to bacteria. Different green tones describe difference of signal producing rate by synthetic cells. (a) Synthetic cells and bacteria are mixed; (b) synthetic cells and bacteria are kept separated. The pink arrows show which are the senders that contribute to deliver the signal molecule to a certain bacterium (Color figure online).

concentration of all species is instantaneously averaged in space, and the positions of the sender and receiver systems are not needed to run the simulations.

On the other hand, it is evident that from the practical viewpoint, different geometries can be realized, as those reported in Fig. 12. Essentially, synthetic cells and bacteria can be either mixed (Fig. 12a) or kept separated (Fig. 12b). In the first case, the two populations are disposed in 3D space as two interpenetrated arrays and the average distance between each synthetic cell and each bacterium is short.

In the second case, the two populations occupy two physically distinct regions of space, and these can be both 3D or 2D/3D (as shown in Fig. 12b). Here, the interface between the two regions is a sort of target surface for the sent molecules S, because the first receivers – those exposed to higher concentration of S – are going to respond efficiently.

A quantitative treatment of signal molecule diffusion in these two cases lies outside the scope of this article since it would require to treat the concentration of signal molecules as dependent both on time and spatial coordinates. In fact, even if vesicles and bacteria are assumed immobile (for example, embedded in a gel [37]) and dimensionless source and sink of signal molecules, respectively, the number of signal molecules that reach a certain target bacterium in a certain time results from the diffusion of such molecules from all synthetic cells present in the medium, and it depends from their spatial distribution, relative distance, and volume density. The model would give information about the response pattern of a population of bacteria exposed to a population of signal-emitting synthetic cells, for example according to the two scenarios proposed in Fig. 12. Based on random or non-random spatial distribution of senders and receivers, these calculations could be helpful to better define the most valuable experimental approaches.

6   Concluding Remarks

Synthetic cell technology, based on a bottom-up SB design, can be adapted and extended for reaching the ambitious goal of interfacing with biological cells, and this is possible by letting synthetic cells share with biological cells a common (chemical) language. These developments are realistic, since synthetic cells, due to their embodied molecular computing capabilities, would share with natural cells many abilities, such as chemical communication, molecular recognition, and signal transduction i.e., the key features for reciprocal interfacing. Here we have presented a mathematical model that simulates the behaviour of synthetic cells sending a chemical signal to biological cells. It is, in essence, a useful tool for the construction of a synthetic-to-natural communication system based on a synthetic biology approach. It will be tested, optimized, modified and refined as soon as experimental data become available. It integrates the wet-lab approach with quantitative evaluations and/or order-of-magnitude estimates, which can be very helpful for designing the experiments in the proper way. The model has two main limitations: the first one is that it is a deterministic model, whereas stochastic effects are pervasive in small-scale systems [31,38]. Next, the spatial information about the diffusion of signal molecules has not been included in this first model. The generation of more realistic models where these to features are taken into account is the challenge for next steps in mathematical modelling synthetic cell/natural cell chemical communication.

More in general, the establishment of a molecular communication technology based on synthetic cells would be a cutting-edge technology for future advancements in practical and theoretical fields. From the practical viewpoint it will pave the way to the construction of soft-wet-micro-robotic systems with implications in nanomedicine and other applications, like bioremediation. From the theoretical viewpoint this will bring to novel strategies for natural computing in the bio-chem-ICTs arena, and possibly establish a novel paradigm for (embodied) artificial intelligence. Moreover, such a tools can be used to investigate basic scientific questions related to the mechanisms of cell signalling taking advantage of operating without the interference of background processes.

Addendum. A report on how synthetic cells can act as "translators" for natural cells has appeared during the preparation of this manuscript. In particular, synthetic cells have been constructed so that they produce α-hemolysin (from internal TX-TL reactions) upon addition of a trigger (theophylline). The consequent formation of a membrane pore (*cf.* with Fig. 4) allows the emission of a previously entrapped molecule (IPTG) that stimulates the bacterial response [39].

Acknowledgements. The in silico model here presented has been recently published, with a more extensive discussion, in [6]. The authors thank Pier Luigi Luisi (Roma Tre University) for inspiring discussions on synthetic minimal cells. F.M. and P.S. acknowledge the support through the COST Action CM1304 (Emergence and Evolution of Complex Chemical Systems).

References

1. Endy, D.: Foundations for engineering biology. Nature **438**, 449–453 (2005)
2. Baldwin, G., Bayer, T., Dickinson, R., Ellis, T., Freemont, P.S., Kitney, R.I., Polizzi, K., Stan, G.B.: Synthetic Biology. A Primer. Imperial College Press, London (2012)
3. Luisi, P.L., Ferri, F., Stano, P.: Approaches to semi-synthetic minimal cells: a review. Naturwissenschaften **93**, 1–13 (2006)
4. Stano, P., Rampioni, G., Carrara, P., Damiano, L., Leoni, L., Luisi, P.L.: Semi-synthetic minimal cells as a tool for biochemical ICT. Biosystems **109**, 24–34 (2012)
5. Rampioni, G., Damiano, L., Messina, M., D'Angelo, F., Leoni, L., Stano, P.: Chemical communication between synthetic and natural cells: a possible experimental design. Electr. Proc. Theor. Comput. Sci. **130**, 14–26 (2013)
6. Rampioni, G., Mavelli, F., Damiano, L., D'Angelo, F., Messina, M., Leoni, L., Stano, P.: A synthetic biology approach to bio-chem-ICT: first moves towards chemical communication between synthetic and natural cells. Nat. Comput. (2014). doi:10.1007/s11047-014-9425-x
7. Nakano, T., Moore, M., Enomoto, A., Suda, T.: Molecular communication technology as a biological ICT. In: Sawai, H. (ed.) Biological Functions for Information and Communication Technologies. Studies in Computational Intelligence, pp. 49–86. Springer, Heidelberg (2011)
8. Amos, M., Dittrich, P., McCaskill, J., Rasmussen, S.: Biological and chemical information technologies. Proc. Comput. Sci. **7**, 56–60 (2011)
9. Nakano, T., Eckford, A.W., Haraguchi, T.: Molecular Communications. Cambridge University Press, Cambridge (2013)
10. Leduc, P.R., Wong, M.S., Ferreira, P.M., Groff, R.E., Haslinger, K., Koonce, M.P., Lee, W.Y., Love, J.C., McCammon, J.A., Monteiro-Riviere, N.A., Rotello, V.M., Rubloff, G.W., Westervelt, R., Yoda, M.: Towards an in vivo biologically inspired nanofactory. Nat. Nanotechnol. **2**, 3–7 (2007)
11. Damiano, L., Hiolle, A., Cañamero, L.: Grounding synthetic knowledge. In: Lenaerts, T., et al. (eds.) Advances in Artificial Life - ECAL 2011, pp. 200–207. MIT Press, Cambridge (2011)
12. Gardner, P.M., Winzer, K., Davis, B.G.: Sugar synthesis in a protocellular model leads to a cell signalling response in bacteria. Nat. Chem. **1**, 377–383 (2009)
13. Stano, P., Kuruma, Y., Souza, T.P., Luisi, P.L.: Biosynthesis of proteins inside liposomes. Methods Mol. Biol. **606**, 127–145 (2010)
14. Stano, P., Carrara, P., Kuruma, Y., Souza, T., Luisi, P.L.: Compartmentalized reactions as a case of soft-matter biotechnology: synthesis of proteins and nucleic acids inside lipid vesicles. J. Mater. Chem. **21**, 18887–18902 (2011)
15. Pautot, S., Frisken, B.J., Weitz, D.A.: Production of unilamellar vesicles using an inverted emulsion. Langmuir **19**, 2870–2879 (2003)
16. Carrara, P., Stano, P., Luisi, P.L.: Giant vesicles "colonies": a model for primitive cell communities. ChemBioChem **13**, 1497–1502 (2012)
17. Grotzky, A., Altamura, E., Adamcik, J., Carrara, P., Stano, P., Mavelli, F., Nauser, T., Mezzenga, R., Schlüter, A.D., Walde, P.: Structure and enzymatic properties of molecular dendronized polymer-enzyme conjugates and their entrapment inside giant vesicles. Langmuir **29**, 10831–10840 (2013)
18. Cabré, E.J., Sánchez-Gorostiaga, A., Carrara, P., Ropero, N., Casanova, M., Palacios, P., Stano, P., Jiménez, M., Rivas, G., Vicente, M.: Bacterial division proteins induce vesicle collapse and cell membrane invagination. J. Biol. Chem. **288**, 26625–26634 (2013)

19. Shimizu, Y., Inoue, A., Tomari, Y., Suzuki, T., Yokogawa, T., Nishikawa, K., Ueda, T.: Cell-free translation reconstituted with purified components. Nat. Biotechnol. **19**, 751–755 (2011)
20. Kuruma, Y., Stano, P., Ueda, T., Luisi, P.L.: A synthetic biology approach to the construction of membrane proteins in semi-synthetic minimal cells. Biochim. Biophys. Acta **1788**, 567–574 (2009)
21. Waters, C.M., Bassler, B.L.: Quorum sensing: cell-to-cell communication in bacteria. Annu. Rev. Cell. Dev. Biol. **21**, 319–346 (2005)
22. West, S.A., Griffin, A.S., Gardner, A., Diggle, S.P.: Social evolution theory for microorganisms. Nat. Rev. Microbiol. **4**, 597–607 (2006)
23. Williams, P., Winzer, K., Chan, W.C., Cámara, M.: Look who's talking: communication and quorum sensing in the bacterial world. Philos. Trans. R. Soc. Lond. Biol. Sci. **362**, 1119–1134 (2007)
24. Atkinson, S., Williams, P.: Quorum sensing and social networking in the microbial world. J. R. Soc. Interf. **6**, 959–978 (2009)
25. Hiyama, S., Moritani, Y., Suda, T., Egashira, R., Enamoto, A., Moore, M., Nakano, T.: Molecular communications. In: Proceedings of the 2005 NSTI Nanotechnology Conference, pp. 391–394 (2005)
26. Noiureaxu, V., Libchaber, A.: A vesicle bioreactor as a step toward an artificial cell assembly. Proc. Natl. Acad. Sci. USA **101**, 17669–17674 (2004)
27. Choi, H.J., Montemagno, C.D.: Artificial organelle: ATP synthesis from cellular mimetic polymersomes. Nano Lett. **5**, 2538–2542 (2005)
28. Paleos, C.M., Tsiourvas, D., Sideratou, Z., Pantos, A.: Formation of artificial multicompartment vesosome and dendrosome as prospected drug and gene delivery carriers. J. Control. Rel. **170**, 141–152 (2013)
29. Chandrawati, R., Caruso, F.: Biomimetic liposome- and polymersome-based multicompartmentalized assemblies. Langmuir **28**, 13798–13807 (2012)
30. Shin, J., Noireaux, V.: An *E. coli* cell-free expression toolbox: application to synthetic gene circuits and artificial cells. ACS Synth. Biol. **1**, 29–41 (2012)
31. Mavelli, F.: Stochastic simulations of minimal cells: the Ribocell model. BMC Bioinf. **13**, S10 (2010)
32. Stano, P., Souza, T., Carrara, P., Altamura, E., D'Aguanno, E., Caputo, M., Luisi, P.L., Mavelli, F.: Recent biophysical issues about the preparation of solute-filled lipid vesicles. Mech. Adv. Mater. Struct. doi:10.1080/15376494.2013.857743
33. Stögbauer, T., Windhager, L., Zimmer, F., Rädler, J.O.: Experiment and mathematical modeling of gene expression dynamics in a cell-free system. Integr. Biol. **4**, 494–501 (2012)
34. Sunami, T., Hosoda, K., Suzuki, H., Matsuura, T., Yomo, T.: Cellular compartment model for exploring the effect of the lipidic membrane on the kinetics of encapsulated biochemical reactions. Langmuir **26**, 8544–8551 (2010)
35. Parsek, M.R., Val, D.L., Hanzelka, B.L., Cronan, J.E., Greenberg, E.P.: Acyl homoserine-lactone quorum-sensing signal generation. Proc. Natl. Acad. Sci. USA **96**, 4360–4365 (1999)
36. Pearson, J.P., Van Delden, C., Iglewski, B.H.: Active efflux and diffusion are involved in transport of *Pseudomonas aeruginosa* cell-to-cell signals. J. Bacteriol. **181**, 1203–1210 (1999)
37. Ullrich, M., Hanuš, J., Dohnal, J., Štěpánek, F.: Encapsulation stability and temperature-dependent release kinetics from hydrogel-immobilised liposomes. J. Colloid Interface Sci. **394**, 380–385 (2013)

38. Mavelli, F., Stano, P.: Kinetic models for autopoietic chemical systems: role of fluctuations in homeostatic regime. Phys. Biol. **7**, 016010 (2010)
39. Lentini, R., Santero, S.P., Chizzolini, F., Cecchi, D., Fontana, J., Marchioretto, M., Del Bianco, C., Terrell, J.L., Spencer, A.C., Martini, L., Forlin, M., Assfalg, M., Dalla Serra, M., Bentley, W.E., Mansy, S.S.: Integrating artificial with natural cells to translate chemical messages that direct *E. coli* behaviour. Nat. Commun. **5**, 4012 (2014)

Examples of the Usage of Infinities and Infinitesimals in Numerical Computations

Yaroslav D. Sergeyev[1,2,3](✉)

[1] University of Calabria, Rende, Italy
[2] N.I. Lobatchevsky State University, Nizhni Novgorod, Russia
[3] Institute of High Performance Computing and Networking, C.N.R., Rende, Italy
yaro@si.dimes.unical.it
http://wwwinfo.dimes.unical.it/~yaro

Abstract. It is well known that traditional computers work numerically with finite numbers only and situations where a use of infinite or infinitesimal quantities is required are studied mainly theoretically by human beings. In this paper, a recently introduced computational methodology that has been proposed with the intention to change this differentiation is discussed. It is based on the principle 'The part is less than the whole' applied to all quantities (finite, infinite, and infinitesimal) and to all sets and processes (finite and infinite). The methodology uses as a computational device the Infinity Computer (patented in USA, EU, and Russian Federation) working numerically with infinite and infinitesimal numbers that can be written using a numeral positional system with an infinite base. On a number of examples it is shown that it becomes possible both to execute computations of a new type and to simplify computations where infinity and/or infinitesimals are required.

Keywords: Numerical infinities and infinitesimals · Infinity computer · Numeral systems

1 Introduction

Even though there exist codes allowing one to work *symbolically* with ∞ and other symbols related to the concepts of infinity and infinitesimals, traditional computers work *numerically* only with finite numbers and situations where the usage of infinite or infinitesimal quantities is required are studied mainly theoretically (see [2,3,6,8,9,12,13,18,19,38] and references given therein). Many among the approaches developed for this purpose are rather old: ancient Greeks following Aristotle distinguished the potential infinity from the actual infinity; John Wallis (see [38]) credited as the person who has introduced the infinity symbol, ∞, has published his work *Arithmetica infinitorum* in 1655; the foundations of analysis we use nowadays have been developed more than 200 years; Georg Cantor (see [2]) has introduced his cardinals and ordinals more than 100 years ago, as well.

© Springer International Publishing Switzerland 2014
C. Pizzuti and G. Spezzano (Eds.): WIVACE 2014, CCIS 445, pp. 190–200, 2014.
DOI: 10.1007/978-3-319-12745-3_15

The fact that numerical manipulations with infinities and infinitesimals have not been implemented so far on computers can be explained by several difficulties. Obviously, among them we can mention the fact that arithmetics developed for this purpose are quite different with respect to the way of computing we use when we deal with finite quantities. For instance, there exist undetermined operations ($\infty - \infty$, $\frac{\infty}{\infty}$, etc.) that are absent when we work with finite numbers. There exist also practical difficulties that preclude an implementation of numerical computations with infinity and infinitesimals. For example, it is not clear how to store an infinite quantity in a finite computer memory.

A new computational methodology introduced recently (see [23,29,33]) allows one to look at infinities and infinitesimals in a new way and to execute *numerical* computations with infinities and infinitesimals on the Infinity Computer patented in USA (see [27]) and other countries. Nowadays there exists a rapidly growing international scientific community that has developed a number of interesting theoretical and applied results using the new methodology in several research areas.

Among them it is worthy to mention studies linking the new approach to the historical panorama of ideas dealing with infinity and infinitesimals (see [14–16,35]). Then, the new methodology has been applied for studying Euclidean and hyperbolic geometry (see [17,20]), percolation (see [10,11,37]), fractals (see [22,24,32,37]), numerical differentiation and optimization (see [4,25,30,40]), infinite series and the Riemann zeta function (see [26,31,39]), the first Hilbert problem and Turing machines (see [28,35,36]), cellular automata (see [5]), ordinary differential equations (see [34]), etc.

In this paper, we briefly describe the new methodology and provide a number of examples showing how it can be used in different situations where infinities and infinitesimals are required. An interested reader is invited to have a look at surveys [23,29,33] and the book [21] written in a popular way.

2 A New Standpoint on Infinity and a New Numeral System

In order to start, let us remind that there exists a distinction (being very important for the new methodology) between *numbers* and *numerals*. A *numeral* is a symbol (or a group of symbols) that represents a *number*. A *number* is a concept that a *numeral* expresses. The same number can be represented by different numerals. For example, the symbols '9', 'nine', 'IIIIIIIII', and 'IX' are different numerals, but they all represent the same number. Rules used to write down numerals together with algorithms for executing arithmetical operations form a *numeral system*.

It is necessary to remind also that different numeral systems can express different sets of numbers and they can be more or less suitable for executing arithmetical operations. Even the powerful positional system is not able to express, e.g., the number π by a finite number of symbols (the finiteness is essential for executing numerical computations) and this special numeral, π, is deliberately introduced to express the desired quantity. There exist many numeral systems that are weaker than the positional one. For instance, Roman numeral system is not able to express zero and negative numbers and such an expression as III–V is an indeterminate form in this numeral system. As a result, before appearing the positional numeral system and inventing zero mathematicians were not able to create theorems involving zero and negative numbers and to execute computations with them. Thus, developing new (more powerful than existing ones) numeral systems can help a lot both in theory and practice of computations.

There exist very weak numeral systems allowing their users to express a very limited quantity of numbers and one of them will be illuminating for our study. This numeral systems is used by a primitive tribe, Pirahã, living in Amazonia nowadays. A study published in *Science* in 2004 (see [7]) describes that these people use an extremely simple numeral system for counting: one, two, many. For Pirahã, all quantities larger than two are just 'many' and such operations as $2 + 2$ and $2 + 1$ give the same result, i.e., 'many'. Using their weak numeral system Pirahã are not able to see, for instance, numbers 3, 4, and 5, to execute arithmetical operations with them, and, in general, to say anything about these numbers because in their language there are neither words nor concepts for that. It is worthy to mention that the result 'many' is not wrong. It is just inaccurate. The introduction of a numeral system having numerals for expressing numbers 3 and 4 leads to a higher accuracy of computations and allows one to distinguish results of operations $2 + 1$ and $2 + 2$.

The weakness of the numeral system of Pirahã leads also to the following results

$$\text{'many'} + 1 = \text{'many'}, \quad \text{'many'} + 2 = \text{'many'}, \quad \text{'many'} + \text{'many'} = \text{'many'}$$

that are crucial for changing our outlook on infinity. In fact, by changing in these relations 'many' with ∞ we get relations used to work with infinity in the traditional calculus

$$\infty + 1 = \infty, \qquad \infty + 2 = \infty, \qquad \infty + \infty = \infty.$$

This comparison suggests that our difficulty in working with infinity is not connected to the nature of infinity but is a result of inadequate numeral systems used to express infinite numbers. In order to increase the accuracy of computations with infinities, the computational methodology developed in [21, 23, 29] proposes a new numeral system that avoids situations similar to 'many'$+1 = $ 'many' providing results ensuring that if a is a numeral written in this numeral system then for any a (i.e., a can be finite, infinite, or infinitesimal) it follows $a + 1 > a$.

The new numeral system works as follows. A new infinite unit of measure expressed by the numeral ① called *grossone* is introduced as the number of elements of the set, $\mathbb{N}$, of natural numbers. Concurrently with the introduction of grossone in the mathematical language all other symbols (like ∞, Cantor's ω, $\aleph_0, \aleph_1, ...$, etc.) traditionally used to deal with infinities and infinitesimals are excluded from the language because ① and other numbers constructed with its help not only can be used instead of all of them but can be used with a higher accuracy. Grossone is introduced by describing its properties postulated by the Infinite Unit Axiom (see [23,29]) added to axioms for real numbers (similarly, in order to pass from the set, $\mathbb{N}$, of natural numbers to the set, $\mathbb{Z}$, of integers a new element – zero expressed by the numeral 0 – is introduced by describing its properties).

The new numeral ① allows us to construct different numerals expressing different infinite and infinitesimal numbers and to execute computations with all of them. As a result, instead of the usual symbol ∞ different infinite and/or infinitesimal numerals can be used. Indeterminate forms are not present and, for example, the following relations hold for infinite numbers ①, $①^2$ and $①^{-1}$, $①^{-2}$ (that are infinitesimals), as for any other (finite, infinite, or infinitesimal) number expressible in the new numeral system

$$0 \cdot ① = ① \cdot 0 = 0, \quad ① - ① = 0, \quad \frac{①}{①} = 1, \quad ①^0 = 1, \quad 1^① = 1, \quad 0^① = 0,$$

$$0 \cdot ①^{-1} = ①^{-1} \cdot 0 = 0, \quad ①^{-1} > 0, \quad ①^{-2} > 0, \quad ①^{-1} - ①^{-1} = 0,$$

$$\frac{①^{-1}}{①^{-1}} = 1, \quad (①^{-1})^0 = 1, \quad ① \cdot ①^{-1} = 1, \quad ① \cdot ①^{-2} = ①^{-1},$$

$$\frac{①^{-2}}{①^{-2}} = 1, \quad \frac{①^2}{①} = ①, \quad \frac{①^{-1}}{①^{-2}} = ①, \quad ①^2 \cdot ①^{-1} = ①, \quad ①^2 \cdot ①^{-2} = 1.$$

The introduction of the numeral ① allows us to represent more infinite and infinitesimal numbers in a unique framework. For this purpose a new numeral system similar to traditional positional numeral systems was introduced in [21,23]. To construct a number C in the numeral positional system with base ①, we subdivide C into groups corresponding to powers of ①:

$$C = c_{p_m} ①^{p_m} + \ldots + c_{p_1} ①^{p_1} + c_{p_0} ①^{p_0} + c_{p_{-1}} ①^{p_{-1}} + \ldots + c_{p_{-k}} ①^{p_{-k}}. \quad (1)$$

Then, the record

$$C = c_{p_m} ①^{p_m} \ldots c_{p_1} ①^{p_1} c_{p_0} ①^{p_0} c_{p_{-1}} ①^{p_{-1}} \ldots c_{p_{-k}} ①^{p_{-k}} \quad (2)$$

represents the number C, where all numerals $c_i \neq 0$, they belong to a traditional numeral system and are called *grossdigits*. They express finite positive or negative numbers and show how many corresponding units $①^{p_i}$ should be added or subtracted in order to form the number C. Note that in order to have a possibility to store C in the computer memory, values k and m should be finite.

Numbers p_i in (2) are sorted in the decreasing order with $p_0 = 0$

$$p_m > p_{m-1} > \ldots > p_1 > p_0 > p_{-1} > \ldots p_{-(k-1)} > p_{-k}.$$

They are called *grosspowers* and they themselves can be written in the form (2). In the record (2), we write $①^{p_i}$ explicitly because in the new numeral positional system the number i in general is not equal to the grosspower p_i. This gives the possibility to write down numerals without indicating grossdigits equal to zero.

The term having $p_0 = 0$ represents the finite part of C since $c_0 ①^0 = c_0$. Terms having finite positive grosspowers represent the simplest infinite parts of C. Analogously, terms having negative finite grosspowers represent the simplest infinitesimal parts of C. For instance, the number $①^{-1} = \frac{1}{①}$ mentioned above is infinitesimal. Note that all infinitesimals are not equal to zero. In particular, $\frac{1}{①} > 0$ since it is a result of division of two positive numbers.

A number represented by a numeral in the form (2) is called *purely finite* if it has neither infinite not infinitesimals parts. For instance, 4 is purely finite and $4 + 3.5①^{-1}$ is not. All grossdigits c_i are supposed to be purely finite. Purely finite numbers are used on traditional computers and for obvious reasons have a special importance for applications.

All of the numbers introduced above can be grosspowers, as well, giving thus a possibility to have various combinations of quantities and to construct terms having a more complex structure. However, in this paper we consider mainly purely finite grosspowers.

Before we conclude this section let us mention that the new numeral system, as all numeral systems, cannot express all numbers and give answers to all questions. Let us consider, for instance, the set of *extended natural numbers* indicated as $\widehat{\mathbb{N}}$ and including $\mathbb{N}$ as a proper subset

$$\widehat{\mathbb{N}} = \{\underbrace{1, 2, \ldots, ① - 1, ①}_{\text{Natural numbers}}, ① + 1, ① + 2, \ldots, ①^2 - 1, ①^2, ①^2 + 1, \ldots\}. \quad (3)$$

What can we say with respect to the number of elements of the set $\widehat{\mathbb{N}}$? The introduced numeral system based on $①$ is too weak to give answers to this question. It is necessary to introduce in a way a more powerful numeral system by defining new numerals (for instance, $②, ③$, etc.).

We finish this section by emphasizing that different numeral systems, if they have different accuracies, cannot be used together. For instance, the usage of '*many*' from the language of Pirahã in the record $4 + $ '*many*' has no any sense because for Pirahã it is not clear what 4 is and for people knowing what 4 is the accuracy of the answer 'many' is too low. Analogously, the records of the type $① + \omega$, $① - \aleph_0$, $①/\infty$, etc. have no sense because they include numerals developed under different methodological assumptions, in different mathematical contests, for different purposes, and, finally, numeral systems these numerals belong to have different accuracies.

3 Examples of Computations with Infinities and Infinitesimals

Let us start by giving an example of multiplication of two infinite numbers A and B written in the numeral system (1), (2) (for a comprehensive description on arithmetical operations see [21, 23]).

Example 1. Let us consider numbers A and B, where

$$A = 2.4①^{45.3} 5.8①^{-7.2}, \qquad B = 6.3①^{5.8} 7.1①^0 9①^{-4.3}.$$

The number A has one infinite part, $2.4①^{45.3}$, and one infinitesimal part equal to $5.8①^{-7.2}$. The number B has one infinite part, $6.3①^{5.8}$, the finite part, 7.1 (remind that $①^0 = 1$), and the infinitesimal one, $9①^{-4.3}$. Their product C is equal to

$$C = B \cdot A = 15.12①^{51.1}17.04①^{45.3}21.6①^{41}36.54①^{-1.4}41.18①^{-7.2}52.2①^{-11.5}.$$

It has three infinite parts and three infinitesimal ones.   □

The new approach gives the possibility to develop a new Analysis (see [26]) where functions assuming not only finite values but also infinite and infinitesimal ones can be studied. For all of them it becomes possible to introduce a new notion of continuity that is closer to our modern physical knowledge. Functions assuming finite and infinite values can be differentiated and integrated.

Example 2. The function $f(x) = x^{2.5}$ has the first derivative $f'(x) = 2.5x^{1.5}$ and both $f(x)$ and $f'(x)$ can be evaluated at infinite and infinitesimal x. Thus, for infinite $x = ①$ we obtain infinite values

$$f(①) = ①^{2.5}, \qquad f'(①) = 2.5①^{1.5}$$

and for infinitesimal $x = 9①^{-1.5}$ we have

$$f(9①^{-1.5}) = 243①^{-3.75}, \qquad f'(9①^{-1.5}) = 67.5①^{-2.25}.$$

Both values, $f(9①^{-1.5})$ and $f'(9①^{-1.5})$, are infinitesimal.   □

We can also work with functions defined by formulae including infinite and infinitesimal numbers.

Example 3. The function $f(x) = \frac{1}{①}x^2 + ①x$ has a quadratic term infinitesimal and the linear one infinite. It has the first derivative $f'(x) = \frac{2}{①}x + ①$. For infinite $x = 4①$ we obtain infinite values

$$f(①) = 4①^2 + 16①, \qquad f'(①) = ① + 8$$

and for infinitesimal $x = ①^{-1}$ we have

$$f(①^{-1}) = 1 + ①^{-3}, \qquad f'(①^{-1}) = ① + 2①^{-2}.$$ □

By using the new numeral system it becomes possible to measure certain infinite sets. As we have seen above, relations of the type 'many' $+ 1 = $ 'many' are consequences of the weakness of numeral systems applied to express numbers. Thus, one of the principles of the new computational methodology consists of adopting the principle 'The part is less than the whole' to all numbers (finite, infinite, and infinitesimal) and to all sets and processes (finite and infinite).

Example 4. Grossone has been introduced in such a way that (see [14, 28, 29] for a detailed discussion) the sets of even and odd numbers have $①/2$ elements each. The set, $\mathbb{Z}$, of integers has $2① + 1$ elements ($①$ positive elements, $①$ negative elements, and zero). Within the countable sets and sets having cardinality of the continuum it becomes possible to distinguish infinite sets having different number of elements expressible in the numeral system using grossone and to see that, for instance,

$$\frac{①}{2} < ① - 1 < ① < ① + 1 < 2① + 1 < 2①^2 - 1 < 2①^2 < 2①^2 + 1 <$$

$$2①^2 + 2 < 2^① - 1 < 2^① < 2^① + 1 < 10^① < ①^① - 1 < ①^① < ①^① + 1. \quad \square$$

In order to see how the principle 'The part is less than the whole' agrees with our traditional views on infinite sets, let us consider the following two examples. The first of them is related to the one-to-one correspondence and takes its origins in studies of Galileo Galilei.

Example 5. The traditional point of view: even numbers can be put in a one-to-one correspondence with all natural numbers in spite of the fact that they are a part of them:

$$
\begin{array}{ll}
\text{even numbers:} & 2,\, 4,\, 6,\, 8,\, 10,\, 12,\, \dots \\
& \updownarrow \updownarrow \updownarrow \updownarrow \updownarrow \updownarrow \\
\text{natural numbers:} & 1,\, 2,\, 3,\, 4 \;\; 5,\;\; 6,\; \dots
\end{array}
\tag{4}
$$

The usual conclusion is that both sets are countable and they have the same cardinality $\aleph_0$. However, now we know that when one executes the operation of counting the accuracy of the result depends on the numeral system used for counting. Since for cardinal numbers it follows

$$\aleph_0 + 1 = \aleph_0, \qquad \aleph_0 + 2 = \aleph_0, \qquad \aleph_0 + \aleph_0 = \aleph_0,$$

these relations suggest that the accuracy of the cardinal numeral system of Alephs is not sufficiently high to see the difference with respect to the number of elements of the two sets.

In order to look at the record (4) using the new numeral system we need the following fact from [21]: ① is even. It is also necessary to remind that numbers that are larger than ① are not natural, they are extended natural numbers. For instance, ① + 2 is even but not natural, it is extended natural, see (3). Since the number of elements of the set of even numbers is equal to $\frac{①}{2}$, we can write down not only initial (as it is usually done traditionally) but also the final part of (4)

$$
\begin{array}{llll}
2,\, 4,\, 6,\, 8,\, 10,\, 12,\, \dots & ① - 4, & ① - 2, & ① \\
\updownarrow \updownarrow \updownarrow \updownarrow \updownarrow \updownarrow & \updownarrow & \updownarrow & \updownarrow \\
1,\, 2,\, 3,\, 4 \;\; 5,\;\; 6,\; \dots & \frac{①}{2} - 2, & \frac{①}{2} - 1, & \frac{①}{2}
\end{array}
\tag{5}
$$

concluding so (4) in a complete accordance with the principle 'The part is less than the whole'. Both records, (4) and (5), are correct but (5) is more accurate, since it allows us to observe the final part of the correspondence that is invisible if (4) is used. $\quad \square$

In order to become more familiar with natural and extended natural numbers we provide one more example.

Example 6. We consider the set of natural numbers, $\mathbb{N}$, and multiply each of its elements by 2. We would like to study the resulting set, that will be called $\mathbb{E}^2$ hereinafter, to calculate the number of its elements and to specify which among its elements are natural and which ones are extended natural numbers and how many they are.

The introduction of the new numeral system allows us to write down the set, $\mathbb{N}$, of natural numbers in the form

$$\mathbb{N} = \{1, 2, \quad \ldots \quad \frac{①}{2} - 2, \frac{①}{2} - 1, \frac{①}{2}, \frac{①}{2} + 1, \frac{①}{2} + 2, \quad \ldots \quad ① - 2, \; ① - 1, \; ①\}.$$

By definition, the number of elements of $\mathbb{N}$ is equal to $①$. Thus, after multiplication of each of the elements of $\mathbb{N}$ by 2, the resulting set, $\mathbb{E}^2$, will also have grossone elements. In particular, the number $\frac{①}{2}$ multiplied by 2 gives us $①$ and $\frac{①}{2} + 1$ multiplied by 2 gives us $① + 2$ that is even extended natural number. Analogously, the last element of $\mathbb{N}$, i.e., $①$, multiplied by 2 gives us $2①$. Thus, the set of even numbers $\mathbb{E}^2$ can be written as follows

$$\mathbb{E}^2 = \{2, 4, 6, \quad \ldots \quad ① - 4, ① - 2, ①, ① + 2, ① + 4, \quad \ldots \quad 2① - 4, 2① - 2, 2①\},$$

where numbers $\{2, 4, 6, \quad \ldots \quad ① - 4, ① - 2, ①\}$ are even and natural (they are $\frac{①}{2}$) and numbers $\{① + 2, ① + 4, \quad \ldots \quad 2① - 4, 2① - 2, 2①\}$ are even and extended natural, they also are $\frac{①}{2}$. $\qquad\square$

The last example is taken from [4]. It is related to the field of nonlinear constrained optimization being an important class of problems with a broad range of scientific and engineering applications. In the literature, there exists a number of algorithms proposed for solving this kind of problems (see, e.g., [1] and references given therein). The authors of [4] concentrate their attention on optimization methods using penalty functions to reduce the original constrained problem to an unconstrained one. Traditionally, this kind of methods requires to solve a sequence of unconstrained minimization problems for increasing values of a penalty parameter and it is necessary to understand what is the point the sequence of solutions converges to. The following example shows that there exist situations where the usage of $①$ as a penalty coefficient allows us to avoid the necessity to solve such a sequence of problems.

Example 7. The authors of [4] consider the following quadratic two-dimensional optimization problem with a single linear constraint

$$\min_{x} \; \frac{1}{2}x_1^2 + \frac{1}{6}x_2^2,$$

subject to $x_1 + x_2 = 1$.

The pair $(\overline{x}, \overline{\pi})$ is a Karush-Kuhn-Tucker point where $\overline{x} = \begin{bmatrix} 1/4 \\ 3/4 \end{bmatrix}$ and $\overline{\pi} = -\frac{1}{4}$.

Then the corresponding unconstrained optimization problem can be constructed using a penalty coefficient P as follows

$$\min_x \ \frac{1}{2}x_1^2 + \frac{1}{6}x_2^2 + \frac{P}{2}(1 - x_1 - x_2)^2.$$

For instance, suppose that we have taken $P = 20$ then the first order optimality conditions can be written as follows

$$\begin{cases} x_1 - 20(1 - x_1 - x_2) = 0, \\ \frac{1}{3}x_2 - 20(1 - x_1 - x_2) = 0. \end{cases}$$

Solution to this system of linear equations is the stationary point of the unconstraint problem, namely, it is

$$x_1^*(20) = \frac{20}{81}, \qquad x_2^*(20) = \frac{60}{81}$$

and it is not clear how to obtain the exact pair $(\overline{x}, \overline{\pi})$ from the pair $(x_1^*(20), x_2^*(20))$. Very often people take a sequence $\{p_k\}$ of increasing values of P, solve the problem again and again, and try to understand where the sequence of points $(x_1^*(p_k), x_2^*(p_k))$ converges.

To avoid the necessity to solve a sequence of problems, the authors of [4] propose to use ① as the penalty coefficient P, i.e., to construct the following unconstrained problem

$$\min_x \ \frac{1}{2}x_1^2 + \frac{1}{6}x_2^2 + \frac{①}{2}(1 - x_1 - x_2)^2.$$

The first order optimality conditions then are

$$\begin{cases} x_1 - ①(1 - x_1 - x_2) = 0 \\ \frac{1}{3}x_2 - ①(1 - x_1 - x_2) = 0 \end{cases}$$

and the solution is

$$x_1^*(①) = \frac{①}{4①+1}, \qquad x_2^*(①) = \frac{3①}{4①+1}.$$

After division we have

$$x_1^*(①) = \frac{1}{4} - ①^{-1}(\frac{1}{16} - \frac{1}{64}①^{-1} + \ldots), \quad x_2^*(①) = \frac{3}{4} - ①^{-1}(\frac{3}{16} - \frac{3}{64}①^{-1} + \ldots).$$

This means that the finite parts of $x_1^*(①)$ and $x_2^*(①)$ give us the exact solution to the original constrained problem. Moreover,

$$-①(1 - x_1^*(①) - x_2^*(①)) = -\frac{1}{4} + \frac{4}{64}①^{-1} + \ldots$$

i.e., we have $\overline{\pi} = -\frac{1}{4}$. $\square$

Thus, the main issue in this example consists of the fact that finite parts of the results can be easily separated from infinitesimals ones. In contrast, in the traditional approaches this is impossible since both the original problem and parameters work with finite values only. Thus, results $(x_1^*(p_k), x_2^*(p_k))$ provided by traditional methods are finite numbers and, therefore, inside the results corresponding to the unconstrained problem one is not able to see the impact of the parameters (perturbations) on the solution of the original problem.

References

1. Boyd, S., Vandenberghe, L.: Convex Optimization. Cambridge University Press, Cambridge (2004)
2. Cantor, G.: Contributions to the Founding of the Theory of Transfinite Numbers. Dover Publications, New York (1955)
3. Conway, J.H., Guy, R.K.: The Book of Numbers. Springer, New York (1996)
4. De Cosmis, S., De Leone, R.: The use of grossone in mathematical programming and operations research. Appl. Math. Comput. **218**(16), 8029–8038 (2012)
5. D'Alotto, L.: Cellular automata using infinite computations. Appl. Math. Comput. **218**(16), 8077–8082 (2012)
6. Gödel, K.: The Consistency of the Continuum-Hypothesis. Princeton University Press, Princeton (1940)
7. Gordon, P.: Numerical cognition without words: Evidence from Amazonia. Science **306**, 496–499 (2004)
8. Hardy, G.H.: Orders of Infinity. Cambridge University Press, Cambridge (1910)
9. Hilbert, D.: Mathematical problems: Lecture delivered before the International Congress of Mathematicians at Paris in 1900. Bull. Am. Math. Soc. **8**, 437–479 (1902)
10. Iudin, D.I., Sergeyev, Y.D., Hayakawa, M.: Interpretation of percolation in terms of infinity computations. Appl. Math. Comput. **218**(16), 8099–8111 (2012)
11. Iudin, D.I., Sergeyev, Y.D., Hayakawa, M.: Infinity computations in percolation theory applications. Commun. Nonlinear Sci. Numer. Simul. **20**(3), 861–870 (2015)
12. Leibniz, G.W., Child, J.M.: The Early Mathematical Manuscripts of Leibniz. Dover Publications, New York (2005)
13. Levi-Civita, T.: Sui numeri transfiniti. Rend. Acc. Lincei (Series 5a) **113**, 7–91 (1898)
14. Lolli, G.: Infinitesimals and infinites in the history of mathematics: A brief survey. Appl. Math. Comput. **218**(16), 7979–7988 (2012)
15. Lolli, G.: Metamathematical investigations on the theory of grossone. Appl. Math. Comput. (2015) (to appear)
16. Margenstern, M.: Using grossone to count the number of elements of infinite sets and the connection with bijections. p-Adic Numbers, Ultrametric. Anal. App. **3**(3), 196–204 (2011)
17. Margenstern, M.: An application of grossone to the study of a family of tilings of the hyperbolic plane. Appl. Math. Comput. **218**(16), 8005–8018 (2012)
18. Newton, I.: Method of Fluxions (1671)
19. Robinson, A.: Non-standard Analysis. Princeton University Press, Princeton (1996)
20. Rosinger, E.E.: Microscopes and telescopes for theoretical physics: How rich locally and large globally is the geometric straight line? Prespacetime J. **2**(4), 601–624 (2011)

21. Sergeyev, Y.D.: Arithmetic of Infinity. Edizioni Orizzonti Meridionali, CS (2003). 2^d electronic edn. (2013)
22. Sergeyev, Y.D.: Blinking fractals and their quantitative analysis using infinite and infinitesimal numbers. Chaos, Solitons Fractals **33**(1), 50–75 (2007)
23. Sergeyev, Y.D.: A new applied approach for executing computations with infinite and infinitesimal quantities. Informatica **19**(4), 567–596 (2008)
24. Sergeyev, Y.D.: Evaluating the exact infinitesimal values of area of Sierpinski's carpet and volume of Menger's sponge. Chaos, Solitons Fractals **42**(5), 3042–3046 (2009)
25. Sergeyev, Y.D.: Numerical computations and mathematical modelling with infinite and infinitesimal numbers. J. Appl. Math. Comput. **29**, 177–195 (2009)
26. Sergeyev, Y.D.: Numerical point of view on Calculus for functions assuming finite, infinite, and infinitesimal values over finite, infinite, and infinitesimal domains. Nonlin. Anal. Ser. A: Theory Methods Appl. **71**(12), e1688–e1707 (2009)
27. Sergeyev, Y.D.: Computer system for storing infinite, infinitesimal, and finite quantities and executing arithmetical operations with them. USA patent 7,860,914 (2010)
28. Sergeyev, Y.D.: Counting systems and the First Hilbert problem. Nonlinear Anal. Ser. A: Theory, Methods Appl. **72**(3–4), 1701–1708 (2010)
29. Sergeyev, Y.D.: Lagrange Lecture: Methodology of numerical computations with infinities and infinitesimals. Rendiconti del Seminario Matematico dell'Università e del Politecnico di Torino **68**(2), 95–113 (2010)
30. Sergeyev, Y.D.: Higher order numerical differentiation on the infinity computer. Optim. Lett. **5**(4), 575–585 (2011)
31. Sergeyev, Y.D.: On accuracy of mathematical languages used to deal with the Riemann zeta function and the Dirichlet eta function. p-Adic Numbers, Ultrametric Anal. Appl. **3**(2), 129–148 (2011)
32. Sergeyev, Y.D.: Using blinking fractals for mathematical modelling of processes of growth in biological systems. Informatica **22**(4), 559–576 (2011)
33. Sergeyev, Y.D.: Numerical computations with infinite and infinitesimal numbers: Theory and applications. In: Sorokin, A., Pardalos, P.M. (eds.) Dynamics of Information Systems: Algorithmic Approaches, pp. 1–66. Springer, New York (2013)
34. Sergeyev, Y.D.: Solving ordinary differential equations by working with infinitesimals numerically on the infinity computer. Appl. Math. Comput. **219**(22), 10668–10681 (2013)
35. Sergeyev, Y.D., Garro, A.: Observability of turing machines: A refinement of the theory of computation. Informatica **21**(3), 425–454 (2010)
36. Sergeyev, Y.D., Garro, A.: Single-tape and multi-tape turing machines through the lens of the Grossone methodology. J. Supercomput. **65**(2), 645–663 (2013)
37. Vita, M.C., De Bartolo, S., Fallico, C., Veltri, M.: Usage of infinitesimals in the Menger's Sponge model of porosity. Appl. Math. Comput. **218**(16), 8187–8196 (2012)
38. Wallis, J.: Arithmetica infinitorum (1656)
39. Zhigljavsky, A.A.: Computing sums of conditionally convergent and divergent series using the concept of grossone. Appl. Math. Comput. **218**(16), 8064–8076 (2012)
40. Žilinskas, A.: On strong homogeneity of two global optimization algorithms based on statistical models of multimodal objective functions. Appl. Math. Comput. **218**(16), 8131–8136 (2012)

The Complex Systems Approach to Protocells

Roberto Serra[1,2](✉)

[1] Department of Physics, Informatics and Mathematics,
Modena and Reggio Emilia University, Modena, Italy
[2] ECLT (European Centre for Living Technology), Venice, Italy
`rserra@unimore.it`

Abstract. Systems biology is mainly focussed upon the description of specific biological items, like for example specific organisms, or specific organs in a class of animals, or specific genetic-metabolic circuits. It therefore leaves open the issue of the search for general principles of biological organization, which apply to all living beings or to at least to broad classes. So the main challenge of complex systems biology is that of looking for general principles in biological systems, in the spirit of complex systems science which searches for similar features and behaviors in various kinds of systems.

I present here some strong arguments in favor of the soundness of this approach, by reviewing data concerning allometric scaling laws and models, focusing in particular on the claim that evolution tends to drive systems towards critical states.

Then I discuss protocells, in particular the very important phenomenon of synchronization between the rate of growth of the proto-genetic material and that of the lipid container, that is a necessary condition for a sustained growth of a population of protocells). Highly simplified, generic models show that such synchronization can be an emergent property under a very wide set of different hypotheses about protocell architectures and kinetic models. Moreover, I discuss the emergence of autocatalytic sets of collectively replicating molecules in a small protocell with a semipermeable membrane, arguing that local differences in the chemical composition of the environment can give rise to a heterogeneous population of protocells.

Keywords: Random boolean networks · Protocells · Critical state · Edge of chaos · Synchronization · Random chemical networks

1 Introduction

Although it is widely agreed that biological systems are complex, there are several important features of the science of complex systems that have not yet deeply affected the study of biological organisms and processes. Indeed, biology has been largely dominated by a gene-centric view in the last decades, and the one gene - one trait approach, which has often proved to be effective, has been

© Springer International Publishing Switzerland 2014
C. Pizzuti and G. Spezzano (Eds.): WIVACE 2014, CCIS 445, pp. 201–211, 2014.
DOI: 10.1007/978-3-319-12745-3_16

extended to cover even complex traits. This simplifying view has been appropriately criticized, and the movement called systems biology [24] has taken off. Systems biology emphasizes the presence of several feedback loops in biological systems, which severely limit the range of validity of explanations based upon linear causal chains (e.g. gene → behavior). Mathematical modeling is one of the favorite tools of systems biologists to analyze the possible effects of competing negative and positive feedback loops which can be observed at several levels (from molecules to organelles, cells, tissues, organs, organisms, ecosystems).

Systems biology is mainly focussed upon the description of specific biological items, like for example specific organisms, or specific organs in a class of animals, or specific genetic-metabolic circuits. It therefore leaves open the issue of the search for general principles of biological organization, which apply to all living beings or to at least to broad classes.

We know indeed that there are some principles of this kind, biological evolution being the most famous one. The theory of cellular organization also qualifies as a general principle. But the main focus of biological research has been that of studying specific cases, with some reluctance to accept (and perhaps a limited interest for) broad generalizations. This may however change, and this is indeed the challenge of complex systems biology[1]: looking for general principles in biological systems, in the spirit of complex systems science which searches for similar features and behaviors in various kinds of systems.

The hope to find such general principles appears well founded, and I will show in Sect. 2 that there are indeed data which provide support to this claim.

Besides data, there are also general ideas and models concerning the way in which biological systems work. The strategy, in this case, is that of introducing simplified models of biological organisms or processes, and to look for their generic properties: this term, borrowed from statistical physics, is used for those properties which are shared by a wide class of systems. In order to model these properties, the most effective approach has been so far that of using ensembles of systems, where each member can be different from another one, and to look for those properties which are widespread. This approach was introduced many years ago by Kauffman (but see [20,21] for a comprehensive discussion) in modeling gene regulatory networks. In Sect. 3, I will describe some of the most important concepts and models of generic properties.

After these brief discussions on data and models that provide evidence in favor of the existence and importance of generic properties, I will focus in Sect. 4 on models of protocells, i.e. simple cell-like structures that might be relevant both for biotechnological purposes and for understanding the origin of life, and I shall show that the complex systems approach to these systems can be particularly interesting, providing useful stimuli to the experimenters.

---

[1] (The term "Complex Systems Biology" was introduced a few years ago by Kaneko [19]. Although it not of widespread use, it seems particularly well suited to indicate an approach to biology which is well rooted in complex systems science.)

2   Generic Properties of Biological Systems: Data

Biologists have been largely concerned with the analysis of specific organisms, and the search for general principles has in a sense lagged behind. This makes sense, since generalizations are hard in biology, however there are also important examples of generic properties (in the sense defined in Sect. 1) of biological systems. Here I will briefly mention only two properties of this kind, namely power-law distributions and scaling laws, which can be observed by analyzing existing data.

Power-law distributions are widespread in biology: for example, the distribution of the activation level of the genes in a cell belongs to this class (see [19] and further references quoted therein). This means that the frequency of occurrence of genes with activation level x, let's call it p(x), is proportional to x^{-g}. Similar laws are found for other important properties, like the abundance of various chemicals in a cell. As it is well-known, power-law distributions differ from the more familiar gaussian distributions in many respects, the most relevant being that there is a higher frequency of occurrence of results which are markedly different from the most frequent ones. These cases, which are present in an appreciable amount, may have a very profound effect on the performance of the system.

It is also well-known that power-law distributions of the number of links are frequently observed in biological networks, like e.g. protein-protein networks or gene networks. In these cases, as well as in many other cases, the power law concerns the distribution of the number of links per node. The remark concerning the relatively high frequency of far-from-average cases applies also here, and this means that there are some "hub" nodes with a very high number of links which most strongly influence the behavior of the network.

Another striking generic property in biology concerns the relationship between the rate of energy consumption (r) and the mass of an organism (m) [41]. We refer here not to single individuals, but to the average values for a given kind of animal (e.g. cow, mouse, hen, etc.). It has been established by several empirical studies that there is a power-law relationship between the rate of oxygen intake (i.e. the energy consumption rate) and the mass, the former being proportional to the mass raised to the exponent 3/4.

Note that although the mathematical relationship is the same in the two cases above, i.e. a power-law, the semantics is very different. In the first example, the power-law refers to a single variable, and to the frequency of occurrence of a given value in a population, while in the second case it refers to the relationship between two different variables.

What is particularly impressive in the relationship between oxygen consumption rate and mass is that it holds for organisms which are very different from each other (e.g. mammals and birds) and that it spans a very wide range of different masses, from whales to unicellular organisms. Moreover, the same relationship can be extrapolated to even smaller masses, and it can be seen that mithocondria and even the molecular complexes involved lay on the same line. So the "law" seems to hold for an astonishingly high range of mass values.

Of course this is not a law stricti sensu, but rather an empirical relationship. It is interesting to observe that an explanation has recently been proposed for this regularity, based on the idea that biological evolution has led different organisms to optimize oxygen use and distribution. Indeed, the value of the exponent, estimated from data, is 3/4, which is surprising, but an elegant proof has been proposed [41] that links the universality of this exponent to the fact that there are three spatial dimensions (and to the hypothesis that evolution works to minimize energy loss).

The two examples discussed above are indeed sufficient to show clearly that there exist generic properties of biological systems, which hold irrespectively of the differences between different organisms.

3   Generic Properties of Biological Systems: Concepts

Several candidate qualitative and quantitative concepts have been proposed to describe the general properties of complex systems, the second principle of thermodynamics being by far the most successful one. In this section I will just briefly mention one of the proposed concepts, that is amenable to at least a partial experimental test, i.e. the notion that evolution should be able to drive biological systems to so-called "critical" states [20–22, 25].

Here "critical" is defined in a specific sense, that is sometimes referred to as "the edge of chaos", and that differs from e.g. the notion of self-organized criticality [1] that has also raised considerable interest in the past. Dissipative deterministic dynamical systems can often show different long-term behaviours, sometimes leading to ordered states (either constant or oscillating in time), sometimes to quite unpredictable, seemingly erratic wanderings in state space. What is more interesting, is that often it is the same dynamical system (defined e.g. by a set of differential equations) that can behave in one way or another, depending upon the values of some parameters. So there are regions in parameter space where the system is ordered, and regions where it is chaotic. Critical states are those that belong to (or, more loosely, that are close to) the boundaries that separate these regions, so they are close to both ordered and chatic states.

It has been suggested [20–22, 25] that critical states provide an optimal trade-off between the need for robustness, since a biological system must be able to keep homeostasis, notwithstanding external as well as internal perturations, and the need to be able to adapt to changes. If this is the case, and if evolution is able to change the network parameters, then it should have driven organisms towards critical regions in parameter space.

This is a very broad and challenging hypothesis, and it can be tested by comparing the results of models of biological systems with data, e.g. models of gene regulatory networks with actual gene expression data. The use of data which is required for this purpose is very different from the more common use of the same data to infer information about the interactions among specific genes. In testing the criticality hypothesis it is instead necessary to look for global properties of gene expression data, like their distributions or some information-theoretic measures.

The model that should be used for comparison should be generic, able to host various dynamical behaviours depending upon the value of some parameter. An outstanding example of this kind of models is that of the Random Boolean Network (RBN) of gene regulatory networks. The expression of a given gene depends upon a set of regulatory molecules, which are themselves the product of other genes, or whose presence is indirectly affected by the expression of other genes. So genes influence each other's expression, and this can be described as a network of interacting genes. In RBNs [20, 21] the activation of a gene is assumed to take just one of two possible values, active (1) or inactive (0) - a boolean approximation whose validity can be judged a posteriori. The model supposes that the state of each node at time t+1 depends upon the values of its inputs nodes at the preceding time step t. Given that the activations are boolean, the function which determines the new state of a node is a boolean function of the inputs.

Searching for generic properties requires consideration of ensembles of networks, that are generated at random (random connections, random boolean functions) while keeping some parameters fixed (e.g., the average number of connections per node). Depending upon the value of a parameter, the families of networks can be ascribed to different dynamical classes, e.g. ordered, critical, disordered.[2]

By comparing experimental data with the properties of ensembles of random networks it is possible to make inferences concerning the values of the parameters that define the set. Recent results [26, 30, 32, 33, 36] indicate that cells like those of the yeast S. cerevisiae and leukemic human cells seem indeed critical. While the conclusion is not definitive, and many more data and analyses are required, these result seem to be very important. They also point to a new way to look for generic properties, which cannot be read directly in the data but can be inferred from a comparison between patterns in data and in model results.

They also provide relevant evidence in favour of the possibility of successfully applying the RBN model to interpret real biological data. Further model improvements have been developed in order to enlarge the set of possible comparisons with experimental tests [15, 16] and the effects of cell-cell interaction in tissues [4, 5, 35, 39].

Finally, it is worth stressing that, by taking into account biological noise, the RBN model has proven able to describe also the main features of cell differentiation [29, 37, 40]: in this way it has been shown that even such a complex phenomenon can be accounted for by a generic model, without the need to introduce ad hoc genetic circuits.[3]

---

[2] For example, *ceteris paribus*, highly connected networks are more disordered than poorly connected ones. This is however a property of the set of networks with those parameter values, and single network realizations can behave in a way different from the typical behavior of their class.

[3] Of course some hypotheses need to be made; in this case, the key hypothesis is that the level of cellular noise is high in stem cells and decreases during differentiation. There are some indications in favour of this hypothesis, which can be subject to experimental testing.

4   Protocell Models

Let us now turn to the contributions that a complex systems approach can provide to the research on protocells. While present day cells are endowed with highly sophisticated mechanisms, which represent the outcome of almost four billion years of evolution, it is generally believed that the first life-forms were much simpler [14,27,28]. Such protocells should have an embodiment structure (micelle or vesicle), a simplified metabolism and a way to give rise to new protocells. Moreover, there should have been a rudimentary genetics, so that the offspring of a cell was "similar" to its parent (at least, more similar on average to a parent than to another randomly chosen protocell).

Besides their interest for the origin of life problem, protocells may be of much practical interest: it is possible to envisage populations of such entities which grow and reproduce, that are selected to perform useful tasks like e.g. drug synthesis.

While protocells have not yet been built[4], it is extremely interesting to understand under which conditions these systems can actually evolve. Models are required to address this issue and, due to the uncertainties about the details, high-level abstract models are particularly relevant. So one is naturally led to consider the properties shared by a large number of models, which may differ under many respects but which are all able to support the basic features of protocells: growth, duplication, inheritance of some traits.

Several different protocell "architectures" have been suggested, most of them based upon lipid vesicles, where an aqueous internal environment is separated from the external water phase by a lipid bilayer, similar to those of existing biological cells. Vesicles form spontaneously under appropriate conditions, and it is known that they are able to split giving rise to two (or more) daughter cells. The different architectures are based on different hypotheses about the chemical composition of the protogenetic material (e.g., nucleic acids, or polypeptides, or even lipids themselves) and about the place where the action, i.e., duplication of genetic molecules and growth of the lipid container, takes place (in the internal environment, in the membrane, at the interface, or some combinations of the two).

One might therefore be tempted to guess that no unified treatment is possible, however this turns out not to be the case: indeed it has been shown that at least one of the major problems in protocell modelling can be dealt with in a unified way. This is the problem of synchronization between the rate of duplication of the lipid container and that of the genetic material; indeed, if such synchronization is not in place and is not stable, sustained growth of a population of protocells is impossible (and of course one is interested in the conditions for this growth, not so much in a single duplication hit).

---

[4] This remark refers to the kind of protocells we are interested in, i.e. those that are built by self-organization and self-assembly starting from various types of molecules, like nucleic acids, polypetides, lipids, etc., avoiding however those that can be obtained only by living beings, like e.g. specialized enzymes. There are other types of entities that are also called protocells, like those that have been obtained by inserting a synthetic genome into a bacterial cell.

Surprisingly, it has been shown that synchronization is an emergent phenomenon that sets spontaneously in while generations follow generations [3,9,23,34].

Moreover, it can be proven to happen in very different protocell architectures, and also under very different hypotheses about the pattern of reactions among the genetic molecules. Synchronization is not always guaranteed, but the conditions under which it takes place have been mathematically characterized and are indeed very broad. It has been observed that even chaotic replicator dynamics can give rise to such synchronization in protocells [13].

Another major question about protocells (and not only protocells) concerns the products of a large web of interacting chemical substances, like e.g. peptides, or nucleotides, or others. They can form polymers of different compositions and different lengths, and a deep theoretical issue concerns the conditions that allow the sustained replication of some of these molecules (indeed, a necessary condition for life).

This problem has a long story and a major reason of interest here is that it has largely been tackled with a "complex systems" orientation, by looking at the generic properties of large webs of reactions and, in order to do so, by considering random reaction networks. It is remarkable that models that differ in their inspiration and mathematical techniques (including those of [6–8,18] all share a common conclusion, i.e. that large collectively autocatalytic sets (ACSs) of molecules should appear, provided that the number of different chemical species exceeds a certain threshold. The formation of each species that belongs to an ACS is catalyzed by at least another species of the ACS.

On the other hand, news from the lab are not so good, so one should try to understand where the difference comes from, avoiding the rough and easy rejection of the model results, on the basis of the simplifications that have been introduced: all models introduce relevant simplifications, and when several of them point in the same direction it is worth trying to understand the reasons why the experiments do not conform to our expectations.

All the "optimistic" theoretical results have been obtained either by using graph-theoretical arguments, or by modelling the reaction kinetics with ordinary differential equations. However, the former approach ignores the dynamics, so it describes relationships among the different types, without taking into account actual concentrations. Concentrations are described in ODE models, but in this case a link that exists between two interacting species continues to exist forever, irrespectively of their actual values. Some phenomenological attempts have been proposed to cure this problem (e.g. setting to zero all the concentrations that fall below a certain level [2] but a more proper account of these chemical processes, that sometimes involve very small numbers of exemplars of a given species, should be inherently stochastic. With my colleagues we simulated one of the best-known random reaction schemes, proposed by Kauffman and others, by using stochastic simulation [10–12]. While I refer to the original papers for further details, it is interesting to observe here that a particular fragility of ACSs has been observed, when the model parameters are such that the random occurrence of cycles is rare but not negligible. In this case, some cycles turn out to be essentially irrelevant because the corresponding reactions are too rare to play a significant role. One might expect that ACSs should be able to avoid this situation, as they are indeed cycles, but they are cycles among catalysts, and some reactions may suffer by lack of substrates. The Kauffman scheme is very elegant in that it does not postulate a priori a distinction between those polymers that are catalysts and those that are substrates, but this brings also the above mentioned problem, that might contribute to explain the difficulty in obtaining ACSs in the lab.

Note that the notion of RAF set [17], that takes into account the food set, i.e. the set of molecules that are supposed be freely available from outside the system, is very useful to understand these aspects, and indeed there is a region of parameter space where ACSs do exist and RAFs do not.

In the end, collectively autocatalytic sets of molecules should be hosted in protocells. A very fundamental question then concerns the possibility that different protocells host different chemical mixtures; if this is the case, then evolution might take place - but if all the protocells have the same chemical make-up, this would not happen. Note also that the most likely scenario should be one where vesicles do form in a homogeneous solution where all the chemical species are already present. In order to explore this possibility, we have developed a very simple protocell model and we have considered the differences in chemical composition due to local fluctuations. Indeed, the small vesicles (whose linear dimensions are about 100 nm) may host on average one or even less than one molecule per type, if the concentration is of micromolar order, therefore there can be important differences in the initial compositions. Preliminary experiments [31] show that these differences might lead to different kinetics and also to different asymptotic states - although further research is of course needed to confirm this claim.

The need to maintain a local chemical environment that differs from the global homogeneous one might actually explain why membranes are at all needed. However, a further reason might be related to a possible active role of the membrane in creating an environment that makes some reactions faster, i.e. playing the role of a catalyst. The same reactions should be accelerated on both sides of the membrane, but if the membrane is semipermeable then the reaction products should quickly dilute in the external environment, while some of them remain more concentrated in the internal part of the prototcell. It is also known that in these conditions the internal concentration of some chemicals may reach much higher values than the external one [38].

In the end, I wish to stress again that the use of generic models of protocells can shed light on very important aspects of their birth, growth and change, and that it can usefully complement the detailed models of specific protocell architectures.

Acknowledgments. I am indebted to my colleague Marco Villani, with whom I shared a 15 years long research experience in complex systems biology, to Stuart Kauffman, for some wonderful discussions, and to my former Ph.D. students (and now post-doc collaborators) Alessandro Filisetti, Alex Graudenzi, Chiara Damiani who made excellent work in exploring the properties of RBNs and of protocell models and in shaping the ideas described here.

References

1. Bak, P., Tang, C., Wiesenfeld, K.: Self-organized criticality. Phys. Rev. A **38**, 364 (1988)
2. Bagley, R.J., Farmer, J.D., Kauffman, S.A., Packard, N.H., Perelson, A.S., Stadnyk, I.M.: Modeling adaptive biological systems. Bio Syst. **23**(2–3), 113–137 (1989)
3. Carletti, T., Serra, R., Poli, I., Villani, M., Filisetti, A.: Sufficient conditions for emergent synchronization in protocell models. J. Theor. Biol. **254**, 741–751 (2008)
4. Damiani, C., Kauffman, S.A., Serra, R., Villani, M., Colacci, A.: Information transfer among coupled random boolean networks. In: Bandini, S., Manzoni, S., Umeo, H., Vizzari, G. (eds.) ACRI 2010. LNCS, vol. 6350, pp. 1–11. Springer, Heidelberg (2010)
5. Damiani, C., Serra, R., Villani, M., Kauffman, S.A., Colacci, A.: Cell-cell interaction and diversity of emergent behaviours. IET Syst. Biol. **5**(2), 137–144 (2011)
6. Dyson, F.J.: Origins of Life. Cambridge University Press, Cambridge (1985)
7. Eigen, M., Schuster, P.: The hypercycle: A principle of natural self-organization. Part A: Emergence of the hypercycle. Die Naturwiss. **64**(11), 541–565 (1977)
8. Farmer, J., Kauffman, S.A., Packard, N.: Autocatalytic replication of polymers. Physica D: Nonlinear Phenom. **220**, 50–67 (1986)
9. Filisetti, A., Serra, R., Carletti, T., Poli, I., Villani, M.: Synchronization phenomena in protocell models. BRL: Biophys. Rev. Lett. **3**(1/2), 325–342 (2008)
10. Filisetti, A., Serra, R., Villani, M., Fuechslin, R., Packard, N., Kauffman, S.A. Poli, I.: A stochastic model of catalytic reaction networks. In: proceedings of ECCS 2010, (European Conference on Complex Systems). CD-Rom Track E, paper p29. ECCS 10 best paper award (2010)
11. Filisetti, A., Graudenzi, A., Serra, R., Villani, M., Fuechslin, R., Packard, N., Kauffman, S.A., Poli, I.: A stochastic model of autocatalytic reaction networks. Theory Biosci. **131**(2), 85–93 (2011)
12. Filisetti, A., Graudenzi, A., Serra, R., Villani, M., De Lucreazia, D., Fuechslin, R., Packard, N., Kauffman, S.A., Poli, I.: A stochastic model of the emergence of autocatalytic cycles. J. Syst. Chem. **2**(2), 2 (2011)
13. Filisetti, A., Serra, R., Carletti, T., Villani, M., Poli, I.: Non-linear protocell models: synchronization and chaos. Europhys. J. B **77**, 249–256 (2010)
14. Ganti, T.: Chemoton Theory (Vol. I. Theory of Fluid Machineries; Vol. II: Theory of Living Systems). KluwerAcademic/Plenum Publishers, New York (2003)
15. Graudenzi, A., Serra, R., Villani, M., Damiani, C., Colacci, A., Kauffman, S.A.: Dynamical properties of a model of gene regulatory network with memory. J. Comput. Biol. **18**(10), 1291–1303 (2011)
16. Graudenzi, A., Serra, R., Villani, M., Damiani, C., Colacci, A., Kauffman, S.A.: Robustness analysis of a model of gene regulatory network with memory. J. Comput. Biol. **18**(4), 559–577 (2011)
17. Hordijk, W., Hein, J., Steel, M.: Autocatalytic Sets and the Origin of Life. Entropy **12**(7), 1733–1742 (2010)
18. Jain, S., Khrishna, S.: Autocatalytic sets and the growth of complexity in an evolutionary model. Phys. Rev. Lett. **81**, 5684–5687 (1998)
19. Kaneko, K.: Complex Systems Biology. Springer, Heidelberg (2006)
20. Kauffman, S.A.: The Origins of Order. Oxford University Press, New York (1993)
21. Kauffman, S.A.: At Home in the Universe. Oxford University Press, New York (1995)

22. Langton, C.G.: Life at the edge of chaos. In: Langton, C.G., Taylor, C., Farmer, J.D., Rasmussen, S. (eds.) Artificial Life II. SFI studies in the sciences of complexity, pp. 41–91. Addison Wesley, Reading (1995)
23. Munteanu, A., Attolini, C.S., Rasmussen, S., Ziock, H., Sole, R.V.: Generic Darwinian selection in protocell assemblies. Phil. Trans. Roy. Soc. Lon. B **362**, 1847 (2007)
24. Noble, D.: The Music of Life. Oxford University Press, Oxford (2006)
25. Packard, N.: Adaptation toward the edge of chaos. In: Kelso, J., Mandell, A., Shle-singer, M. (eds.) Dynamic Patterns in Complex Systems, pp. 293–301. World Scientific, Singapore (1988)
26. Ramo, P., Kesseli, Y., Yli-Harja, O.: Perturbation avalanches and criticality in gene regulatory networks. J. Theor. Biol. **242**, 164 (2006)
27. Rasmussen, S., Chen, L., Deamer, D., Krakauer, D.C., Packard, N.H., Stadler, P.F., Bedeau, M.A.: Transitions from non living to living matter. Science **303**, 963–965 (2004)
28. Rasmussen, S., Bedeau, M.A., Chen, L., Deamer, D., Krakauer, D.C., Packard, N.H., Stadler, P.F.: Protocells. MIT Press, Cambridge (2009)
29. Ribeiro, A.S., Kauffman, S.A.: Noisy attractors and ergodic sets in models of gene regulatory networks. J. Theor. Biol. **247**, 743–755 (2007)
30. Shmulevich, I., Kauffman, S.A., Aldana, M.: Eukaryotic cells are dynamically ordered or critical but not chaotic. PNAS **102**, 13439–13444 (2005)
31. Serra, R., Filisetti, A., Villani, M., Graudenzi, A., Damiani, C., Panini, T.: A stochastic model of catalytic reaction networks in protocells. Nat. Comput. **13**, 367–377 (2014)
32. Serra, R., Villani, M., Semeria, A.: Genetic network models and statistical properties of gene expression data in knock-out experiments. J. Theor. Biol. **227**, 149–157 (2004)
33. Serra, R., Villani, M., Graudenzi, A., Kauffman, S.A.: Why a simple model of genetic regulatory networks describes the distribution of avalanches in gene expression data. J. Theor. Biol. **246**, 449–460 (2007)
34. Serra, R., Carletti, T., Poli, I.: Synchronization phenomena in surface reaction models of protocells. Artif. Life **13**, 1–16 (2007)
35. Serra, R., Villani, M., Damiani, C., Graudenzi, A., Colacci, A.: The diffusion of perturbations in a model of coupled random boolean networks. In: Umeo, H., Morishita, S., Nishinari, K., Komatsuzaki, T., Bandini, S. (eds.) ACRI 2008. LNCS, vol. 5191, pp. 315–322. Springer, Heidelberg (2008)
36. Serra, R., Villani, M., Graudenzi, A., Colacci, A., Kauffman, S.A.: The simulation of gene knock-out in scale-free random boolean models of genetic networks. Netw. Heterogen. Media **3**(2), 333–343 (2008)
37. Serra, R., Villani, M., Barbieri, A., Kauffman, S.A., Colacci, A.: On the dynamics of random Boolean networks subject to noise: attractors, ergodic sets and cell types. J. Theor. Biol. **265**, 185–193 (2010)
38. Serra, R., Villani, M.: Mechanism for the formation of density gradients through semipermeable membranes. Phys. Rev. E **87**(6), 062814 (2013)
39. Villani, M., Serra, R., Ingrami, P., Kauffman, S.A.: Coupled random boolean network forming an artificial tissue. In: El Yacoubi, S., Chopard, B., Bandini, S. (eds.) ACRI 2006. LNCS, vol. 4173, pp. 548–556. Springer, Heidelberg (2006)

40. Villani, M., Barbieri, A., Serra, R.: A dynamical model of genetic networks for cell differentiation. PLoS ONE **6**(3), e17703 (2011)
41. West, G.B., Brown, J.H.: The origin of allometric scaling laws in biology from genomes to ecosystems: towards a quantitative unifying theory of biological structure and organization. J. Exper. Biol. **208**, 1575 (2005)

Author Index